MW01630570

Faculty Guide

TO THE TELECOURSE

Unseen Life on Earth:
An Introduction to
Microbiology

WITHDRAWN

Faculty Guide

TO THE TELECOURSE
Unseen Life on Earth: An Introduction to Microbiology

to accompany

MICROBIOLOGY

Fourth Edition

Lansing M. Prescott
Augustana College

John P. Harley
Eastern Kentucky University

Donald A. Klein
Colorado State University

Prepared by
John P. Harley
Eastern Kentucky University

Boston   Burr Ridge, IL   Dubuque, IA   Madison, WI   New York   San Francisco   St. Louis
Bangkok   Bogotá   Caracas   Lisbon   London   Madrid
Mexico City   Milan   New Delhi   Seoul   Singapore   Sydney   Taipei   Toronto

McGraw-Hill Higher Education

A Division of The McGraw·Hill Companies

FACULTY GUIDE TO THE TELECOURSE UNSEEN LIFE ON EARTH:
AN INTRODUCTION TO MICROBIOLOGY

Copyright © 2000 by The McGraw-Hill Companies, Inc. and Corporation for Public Broadcasting. All rights reserved. Printed in the United States of America. Except as permitted under the United States Copyright Act of 1976, no part of this publication may be reproduced or distributed in any form or by any means, or stored in a data base or retrieval system, without the prior written permission of the publisher.

This book is printed on acid-free paper.

1 2 3 4 5 6 7 8 9 0 QPD/QPD 9 0 9 8 7 6 5 4 3 2 1 0 9

ISBN 0–07–237428–4

Vice president and editorial director: *Kevin T. Kane*
Publisher: *James M. Smith*
Developmental editor: *Jean Sims Fornango*
Senior marketing manager: *Martin J. Lange*
Senior marketing assistant: *Tami Petsche*
Project manager: *Vicki Krug*
Production supervisor: *Enboge Chong*
Designer: *K. Wayne Harms*
Senior photo research coordinator: *Carrie K. Burger*
Compositor: *Carlisle Communications, Ltd.*
Typeface: *11/12 Times*
Printer: *Quebecor Printing Book Group/Dubuque, IA*

Cover designer: *Kathy Theis*

www.mhhe.com

CONTENTS

PREFACE

THE UNSEEN LIFE ON EARTH: AN INTRODUCTION TO MICROBIOLOGY *Telecourse Faculty Guide*, offers a plan for college and universities that license UNSEEN LIFE ON EARTH as a complete college-credit course.

Part I: Introductory Material contains information designed to help both faculty who are new to telecourse teaching and experienced professors. It also provides specific advise for those who will administer the telecourse.   The *Telecourse Faculty Guide* gives faculty all the support they need to effectively plan a distance learning course in microbiology and the traditional college and high school classroom. In addition to general microbiology, the *Telecourse Faculty Guide* can also be used for courses in biology, life sciences, as well as for staff training in microbial-related industries.

Part II: Guide for Individual Programs contains a set of teaching resources for each of the twelve video programs that comprise **UNSEEN LIFE ON EARTH: AN INTRODUCTION TO MICROBIOLOGY**. These resources are offered as information and suggestions to complement the range of experience that each individual instructor brings to the course. While these ideas may serve to stimulate, they by no means exhaust the many creative teaching strategies that will occur to teachers as they prepare to teach microbiology to their distant learning students. The teaching resources for each video program includes these elements:

Quote and Significance of the Quote: The American Society for Microbiology has recently published "Microbiology's fifty most significant events during the past 125 years" in order to show the importance of microbiology to our everyday lives. Quotes are a way of getting one's immediate attention and are designed to perk student interest and provide perspective on the programs content. These two devices set the stage for what is to follow.

Student Assignments: The student is directed to the specific video to watch, textbook chapter(s) in *Microbiology*, 4th ed., Prescott/Harley/Klein, McGraw-Hill to read, and suggested laboratory exercises from *Laboratory Exercises in Microbiology*, 4th ed., Harley/Prescott, McGraw-Hill.

Lesson Overview: This section provides an integrated summary of information, ideas, and examples presented in the video and assigned chapters for the video. Several statements briefly summarize some of the most important concepts the students should master.

Learning Objectives: A list of objectives is presented for each program, based upon the material covered in the video and chapters.

Course Integrator: This part of the guide ties it all together. It integrates all of the products offered by McGraw-Hill for *Microbiology* 4th edition. Examples include *Microbes in Motion II* CD-ROM; *Hyperclinic: Interactive Case Studies in Microbiology* CD-ROM; *Indentibacter CD-ROM*; *Visual Resource Library* CD-ROM; specific web sites; transparencies; slides; Student Online Learning Center (includes: reviews, quizzes, internet links, microbiology news, microbiology careers); and Instructor Online Learning center (includes: teaching related web links and microbiology news).

Microbial Terms Used in the Video in Order of Appearance with Page Indexing to the Textbook *Microbiology*, 4th ed., Prescott/Harley/Klein, McGraw-Hill: A list of all microbiology key terms is provided to emphasize the most significant facts and concepts. Each term is page-referenced to the page on which the term is first introduced in the textbook chapter.

Web Links Related to the Video and Readings in the Chapter(s): These can be accessed directly from www.mhhe.com/prescott or by specific URLs listed for each link.

Microbiology Terms Used in the Chapter(s): A list of all boldfaced terms is provided for each textbook chapter to emphasize the most significant facts and concepts. Each term is page-referenced to the page on which the term is first introduced in the chapter.

Sample Test Questions from the Video,
Sample Test Questions from Text Chapter(s) Web Sites, and
Sample Test Questions from the Chapter(s): Multiple choice, true or false, matching, labeling, and critical thinking questions are provided specifically from the above sources.

Answers to the Above Sets of Questions

CURRICULUM RECOMMENDATIONS FOR INTRODUCTORY MICROBIOLOGY COURSES AS PREPARED BY THE *AMERICAN SOCIETY FOR MICROBIOLOGY* (ASM)

Since 1994, faculty from institutions across the continent and across the range from community colleges through research universities defined the common ground for *all* introductory microbiology courses. This includes the recommendations below and endorsement of a required laboratory experience. This inclusion of a laboratory experience as an integral part of a microbiology course has been reaffirmed each year at the annual, undergraduate microbiology conference sponsored by the American Society for Microbiology (ASM). The annual conferences have fostered teaching practices to enhance learning based on these guidelines and have led to the development of curriculum materials including the telecourse, *Unseen Life on earth: An Introduction to Microbiology.*

ASM recommends the following core curriculum guidelines for all introductory microbiology courses. These include prerequisite courses, core themes and concepts, and a core laboratory curriculum. The telecourse, *Unseen Life on Earth: An Introduction to Microbiology*, supports these core curriculum recommendations. Updates to the core curriculum recommendations can be found at the ASM website at www.asmusa.org.

Prerequisite Courses. Introductory microbiology courses assume that students have acquired essentail knowledge and skills from introductory biology and introductory chemistry courses at the undergraduate level.

Core Themes and Concepts. The core themes and concepts, considered essential to teach in every introductory microbiology laboratory regardless of its emphasis, includes five overarching themes and twenty essential concepts. Faculty might add items appropriate to allied health, applied environmental, or microbiology-major courses.

The core themes and concepts are not meant to be a syllabus or outline. Rather, these themes and concepts are meant to support the development of learning objectives that can be met within the introductory microbiology course.

In these recommendations, the term "microbes" refers to all microorganisms whether they are subcellular viruses and other infectious agents or cellular parts including all prokaryotic and eukaryotic microbes. The asterisk denotes those themes and concepts considered essential to the laboratory content core.

Theme 1: Microbial Cell Biology*
1. Information flow within a cell
2. Regulation of cellular activities
3. Cellular structure and function*
4. Growth and division*
5. Energy cell metabolism*

Theme 2: Microbial Genetics*
1. Inheritance of genetic information
2. Causes, consequences and uses of mutations*
3. Exchange and acquisition of genetic information

Theme 3: Interactions and Impact of Microorganisms and Humans*
1. Host defense mechanisms
2. Microbial pathogenicity mechanisms*
3. Disease transmission
4. Antibiotics and chemotherapy*
5. Genetic engineering
6. Biotechnology

Theme 4: Interactions and Impact of Microorganisms in the Environment*
1. Adaptation and natural selection
2. Symbiosis
3. Microbial recycling of resources
4. Microbes transforming environment

Theme 5: Integrating Themes*
1. Microbial evolution
2. Microbial diversity*

Laboratory Core Curriculum. The laboratory core curriculum, considered essential to teach in every introductory microbiology laboratory regardless of its emphasis, includes content themes (section above marked with an asterisk), laboratory skills, laboratory thinking skills, and safety. Faculty might add items appropriate to allied health, applied, environmental, or microbiology-major courses.

The laboratory core is not meant to be a syllabus or outline. Rather, these core skills and topics are meant to support the development of learning objectives that can be met within the introductory microbiology laboratory. Depending on the specific emphasis of the course, a single laboratory session could meet multiple core objectives, focus on one objective, or emphasize a topic that is not in the lab core but is important to that particular course.

Laboratory Skills

A student successfully completing basic microbiology will demonstrate the ability to

1. Use a bright-field light microscope to view and interpret slides, including
    a.   Correctly setting up and focusing the microscope
    b.   Proper handling, cleaning, and storage of the microscope
    c.   Correctly using all lenses
    d.   Recording microscopic observations

2. Properly prepare slides for microbiological examination, including
    a.   Cleaning and disposing of slides
    b.   Preparing smears from solid and liquid cultures
    c.   Performing wet mount and/or hanging drop preparations
    d.   Performing Gram stains

3. Properly use aseptic techniques for the transfer and handling of microorganisms and instruments, including
    a.   Sterilizing and maintaining sterility of transfer instruments
    b.   Performing aseptic transfer
    c.   Obtaining microbial samples

4. Use appropriate microbiological media and test systems, including
    a.   Isolating colonies and/or plaques
    b.   Maintaining pure cultures
    c.   Using biochemical test media
    d.   Accurately recording macroscopic observations

5. Estimate the number of microbes in a sample using serial dilution techniques, including
    a.   Correctly choosing and using pipettes and pipetting devises
    b.   Correctly spreading diluted samples for counting
    c.   Estimating appropriate dilutions
    d.   Extrapolating plate counts to obtain the correct colony forming units (CFU) or plaque forming units (PFU) in the starting sample

6. Use standard microbiology laboratory equipment correctly, including
    a.   Using the standard metric system for weights, lengths, diameters, and volumes
    b.   Lighting and adjusting a laboratory burner
    c.   Using an incubator

Laboratory Thinking Skills

A student successfully completing basic microbiology will demonstrate an increased skill level in

1. Cognitive processes, including
    a. Formulating a clear, answerable question
    b. Developing a testing hypothesis
    c. Predicting expected results
    d. Following an experimental protocol

2. Analysis skills, including
    a. Collecting and organizing data in a systematic fashion
    b. Presenting data in an appropriate form (graphs, tables, figures, or descriptive paragraphs)
    c. Assessing the validity of the data (including integrity and significance)
    d. Drawing appropriate conclusions based on the results

3. Communication skills, including
    a. Discussing and presenting lab results or findings in the laboratory

4. Interpersonal and citizenry skills, including
    a. Working effectively in teams or groups so that the task, results, and analysis are shared
    b. Effectively managing time and tasks allowing concurrent and/or overlapping tasks to be done simultaneously, by individuals and within a group
    c. Integrating knowledge and making informed judgments about microbiology in everyday life

Laboratory Safety

A student successfully completing basic microbiology will demonstrate the ability to explain and practice safe

1. Microbiological procedures, including
    a. Reporting all spills and broken glassware to the instructor and receiving instructions for clean up
    b. Using methods for aseptic transfer
    c. Minimizing or containing the production of aerosols and describing the hazards associated with aerosols
    d. Washing hands prior to and following laboratories and at any time contamination is suspected
    e. Using universal precautions with blood and other body fluids and following the requirements of OSHA Bloodborne Pathogen Standard
    f. Disinfecting lab benches and equipment prior to and at the conclusion of each lab session, using an appropriate disinfectant and allowing a suitable contact time
    g. Identifying and disposing properly different types of waste
    h. Reading and signing a laboratory safety agreement indicating that the student has read and understands the safety rules of the laboratory
    i. Following good laboratory practices, including returning materials to proper locations, proper care and handling of equipment, and keeping the bench top clear of extraneous material

2. Protective procedures, including

    a. Tying long hair back, wearing protective equipment (eye protection, coats, gloves, closed shoes; glasses may be preferred to contact lenses), and using such equipment in appropriate situations

    b. Always using appropriate pipetting devises and understanding that mouth pipetting is forbidden

    c. Never eating or drinking in the laboratory

    d. Never applying cosmetics, handling contact lenses, or placing objects (fingers, pencils) in the mouth or touching the face

3. Emergency procedures, including

    a. Locating and properly using emergency equipment (eye wash stations, first aid kits, fire extinguishers, chemical safety showers, telephones, and emergency numbers)

    b. Reporting all injuries immediately to the instructor

    c. Following proper steps in the event of an emergency

In addition, institutions where microbiology laboratories are taught will

    a. Train faculty and staff in proper waste management

    b. Provide and maintain all necessary safety equipment and information resources

    c. Train faculty, staff and students in the use of safety equipment and procedures

    d. Train faculty and staff in the use of MSDS

Correlation Guide to the ASM Core Curriculum

This correlation guide was compiled in order to assist you in integrating the textual material presented in the fourth edition of *Microbiology* by Prescott/Harley/Klein (McGraw-Hill) into the Undergraduate Microbiology Curriculum recommended by the American Society for Microbiology (*see pages viii-xi*).

The American Society for Microbiology has recommended a core curriculum designed as a platform on which microbiology should be taught. The curriculum was developed by faculty from community colleges, four-year undergraduate colleges, and universities to serve as a common experience for all students taking an initial microbiology course. The recommendation also includes a core laboratory curriculum to augment and extend the lecture material. The ASM currently preparing a peer-reviewed database of curricular material for both the lecture and laboratory components of a microbiology course. These content-rich exercises will provide a multi-faceted experience for microbiology students.

ASM Theme	*Microbiology*, 4th ed., Prescott/Harley/Klein, McGraw-Hill
Microbial Cell Biology	
Information flow within a cell	pp 212-215
Regulation of cellular activities	pp 237-252
Cellular structure and function	
Prokaryotes	pp 37-69; 422-520
Eukaryotes	pp 73-93; 524-526; 542-543; 555-556
Viruses	pp 342-353; 363-364; 373; 384; 387-389
Growth and Division	
Prokaryotes	pp 114-132
Eukaryotes	pp 528-535; 543-544; 556
Viruses	pp 336-342; 358-370; 373; 383; 385-386
Cell energy and metabolism	
Prokaryotes	pp 98-110; 152-208
Eukaryotes	pp 527; 543; 556
Viruses	pp 337

Microbial Genetics

Inheritance of genetic information	pp 256-262; 279-281
Causes, consequences and uses of mutations	pp 262-274
Exchange and acquisition of genetic information	pp 281-307

Interactions and Impact on Humans

Host defense mechanisms	pp 591-602; 606-674
Pathogenicity mechanisms	pp 581-591
Disease transmission	pp 722-735
Antibiotics and chemotherapy	pp 678-695
Genetic engineering	pp 311-332
Biotechnology	pp 932-960

Interactions and Impact on the Environment

Adaptation and natural selection	pp 245; 395-396
Symbiosis	pp 566-578
Recycling of resources	pp 245; 395-396
Transforming the environment	pp 908-960

Integrating Themes

Evolution	pp 245; 395-396
Diversity	pp 13; 394-413

Web Resources for Microbiology Instructors

The following Web Resources can be accessed via the American Society for Microbiology:
ASM's Active Learning for Microbiology Classrooms and Labs (*www.asmusa.org/division/w*) hot links.

ASM's Selected References for Microbiology Educators.

On-line Educator: making the Internet an accessible, useful classroom tool. Updated each Monday
morning. Subject material mainly for K-12 instructors, however "Tools for Teachers" and other subjects
are relevant for all instructors.

Biology Education Review : A newsletter intended to encourage communication and the sharing of ideas
between college biology teachers. Recent features include Using News Media in Biology Education. Good
list of journals for biology educators.

Case Studies in Science. State University of New York, Buffalo. "Although the case method has been used
for years to teach law, business, and medicine, it is not common in science. Yet the use of case studies
holds great promise as a pedagogical technique for teaching science, particularly to undergraduates,
because it humanizes science and well illustrates scientific methodology and values. It develops students'
skills in group learning, speaking, and critical thinking, and since many of the best cases are based on
contemporary--and often contentious--science problems that students encounter in the news (such as human
cloning), the use of cases in the classroom makes science relevant."

Access Excellence: A place in cyberspace for Biology Teaching and Learning. Access Excellence is a
national educational program sponsored by Genentech, Inc.

Penn State Altoona. Microbiology Web Page for Undergraduate Research. Features include Research
Sites/Undergraduate Mentoring Network /Sites of Interest to Teachers of Microbiology /Use of Art in
Teaching Microbiology /Sites Related to Lab Safety /Other Sites of Interest to Microbiologists /Biofilm
Related Activities

Sites Related to Laboratory Safety: (Compiled by Penn State Altoona). "The Web is a rich source of
information on safety practices in the laboratory. The following are a few sites that have been found useful
in training students in safe laboratory practice. Also included are sites at which Material Safety Data
Sheets can be viewed and downloaded."

Medical Microbiology Textbook: This website contains the entire Medical Microbiology textbook by
Baron. The textbook includes all the images and figures in the printed form of the text. A wonderful
resource for those teaching medical microbiology.

Demonstration site for the Instructional Library, sponsored by the American Society for Microbiology.
"The Demonstration site includes a variety of curricular resources: an image collection with animations,
videos, and still images of microbes; and innovative laboratory and lecture activities. The site has been
developed for educators, students, and prospective authors who are interested in submitting materials to
this international collection. The fully searchable library will be on-line in June 1999"

More Microbiology Bookmarks for Instructors
(Compiled by Chris Woolverton, Kent State University, Kent, Ohio)

Safety Resources
Comprehensive Safety Manual   *http://www.llnl.gov/es_and_h/hsm/llnl_hc.shtml*

Online links to MSDS information   *http://www.ehs.psu.edu/msdsreq.html*

Penn State University Biosafety Program   *http://www.ehs.psu.edu/mainmap.map?68*

American Conference of Governmental Industrial Hygienists, Inc. *http://www.acgih.org/links/welcome.htm*

CDC Online Biosafety Manuals   *http://www.cdc.gov/od/ohs/manual/mannav.htm*

Laboratory Safety Program (guidelines and practices)   *http://www.ehs.psu.edu/labsafety.html*

Searchable OSHA standards   *http://www.osha-slc.gov/OshStd_toc/OSHA_Std_toc.html*

Agency for Toxic Substances and Disease Registry Searchable HazDat Database
http://atsdr1.atsdr.cdc.gov:8080/hazdat.html#A3.1.2a

National Institute for Occupational Safety and Health   *http://www.cdc.gov/niosh/homepage.html*

Free Access to search ToxLine   *http://www.medscape.com/misc/FormToxlineInfLive.html*

Free Access to search MedLine   *http://www.ncbi.nlm.nih.gov/PubMed/*

National Institute of Environmental Health Sciences   *http://www.niehs.nih.gov/*

US EPA Searchable site   *http://www.epa.gov/*

Online Reporting of Biological and Chemical Outbreaks
http://www.outbreak.org/cgi-unreg/dynaserve.exe/cb/index.html

Teaching Resources
Harding Meta Directory of Microbiology and Infectious Diseases
http://www.lib.uiowa.edu/hardin/md/micro.html

BioChemNet   *http://schmidel.com/bionet.htm*

CDC Homepage   *http://www.cdc.gov/*

Genetic Engineering News   *http://www.genengnews.com/*

U.S. National Library of Medicine   *http://www.nlm.nih.gov*

National Center for Infectious Diseases *http://www.cdc.gov/ncidod/diseases*

American Society for Microbiology *http://www.asmusa.org/*

American Society for Microbiology --Microbiology Education *http://www.asmusa.org/division/w/*

Infectious Disease News (Slack Inc.) *http://www.slackinc.com/general/idn/idnhome.htm*

The Pasteur Institute *http://www.pasteur.fr/welcome-uk.html*

Library of Congress World Wide Web Home Page *http://www.loc.gov/*

Microbiology Jump Station *http://www.horizonpress.com/gateway/micro.html*

Molecular Biology Gateway *http://www.horizonpress.com/gateway/*

Molecular Biology Protocols Gateway *http://www.horizonpress.com/gateway/protocols.html*

NAMES Project *http://www.aidsquilt.org*

National Foundation for Infectious Diseases *http://www.nfid.org*

National Institute of Allergy and Infectious Disease *http://www.niaid.nih.gov/*

National Institutes of Health *http://www.nih.gov/*

Community of Science *http://www.cos.com/*

Harvard University Molecular Biology Gateway *http://golgi.harvard.edu/*

World Health Organization WWW Home Page *http://www.who.int/*

Gram-staining Procedure *http://medic.med.uth.tmc.edu/path/grampro.htm*

Bug-Bytes at LSU *http://www.ccm.lsumc.edu:80/bugbytes/*

Mycology Teaching Resources *http://www.keil.ukans.edu/~fungi/fteach.html*

Bock Laboratories at UW-Madison (Virology, Molecular Biology & Microscopy)
http://www.bocklabs.wisc.edu

Bugs in the News *http://falcon.cc.ukans.edu/~jbrown/bugs.html*

Classic Papers in Genetics *http://www.esp.org/foundations/genetics/classical/*

The Microbe Files (WSU Microbiology) *http://www.wsu.edu:8080/~hurlbert/*

Web related Immunology Class Resources   *http://www.umass.edu/microbio/immclass/*

Digital Learning Center's Microbes in the News
http://commtechlab.msu.edu/CTLProjects/dlc-me/news/news.html

The Microbial Underground's Guide to Microbiology on the Net (comprehensive)
http://www.lsumc.edu/campus/micr/mirror/public_html/microbio.html

Molecular Expressions: Images from the Microscope   *http://micro.magnet.fsu.edu/*

Morbidity and Mortality Weekly Report   *http://www.cdc.gov/epo/mmwr/mmwr.html*

National Science Foundation   *http://stis.nsf.gov/*

The Education & Career Center *http://www.peters* ons.com

Harvard University Gateway to Immunology Resources   *h ttp://golgi.harvard.edu/biopages/immuno.html*

Rotavirus Online http://rotavirus.com/

PART I: INTRODUCTORY MATERIAL
(OFFERING A DISTANCE LEARNING TELECOURSE)[1]

Introduction:
Good planning – well in advance of the first broadcast or viewing, by a team of faculty, administrators, and support personnel – is the key to presenting a good microbiology telecourse. This portion of the *Telecourse Faculty Guide* will serve as an aid to your planning. It suggests ways to make your college's offering of *Unseen Life on Earth: An Introduction to Microbiology* telecourse work smoothly and effectively.

The ideas and strategies presented here are based on the experience of other institutions that have successfully offered telecourses. This guide owes its text to the suggestions of many adult learning experts. Since every educational system and every institution within it has an individual style of organization, these guidelines are meant to be suggestions, not rules.

About Telecourses:
A telecourse is a complete and integrated instructional system that generally includes television programs, a text, student study guide, and a variety of other instructional materials designed to be used by faculty and student as the principal components of a college course. The materials generally reflect the contributions of top scholars and curriculum designers from throughout the country and have been evaluated and refined.

Although enrollment in telecourses varies, in many of the courses 60 to 65 percent of the students are women. Students find that the courses provide a structure for the efficient use of time, allowing them to keep up regular commitments in the workplace as well as at home. When surveyed, many say that they would not have enrolled in a standard class on a given subject if the telecourse option had not been offered. Students tend to busy, working adults who choose telecourses because they reduce travel costs and time, and permit a heavier course load without a proportionate increase in commuting or class time.

Telecourses are, in most other respects (e.g., academic rigor, student requirements, and need for qualified faculty), equivalent to traditionally taught college courses. Telecourse students enroll in the college or university that has adopted the course, pay tuition to that institution, usually have access to all student services offered by that institution, receive academic credit from that institution, and study under the rules and regulations of that institution. Faculty assigned by the adopting institutions are responsible for their students' progress and evaluation and make academic and instructional decisions similar to those for traditionally taught classes. Faculty and students interact through a combination of meetings, written assignments, phone contact, e-mail, course web sites maintained by the instructor, and regular mailings.

[1] Courtesy of Kendall Hunt Publishing, adapted from *Exploring the World of Music Faculty Guide*, Diane U. Eisenberg. Used by permission.

Guidelines for Faculty:

Prepare Yourself
Being a telecourse instructor requires as much careful preparation as any other teaching assignment, although your responsibilities will not be exactly the same as for your on-campus classes. The amount of personal contact between student and faculty will differ according to how your responsibilities are defined, how the course is coordinated, how much you shape the course to reflect your own area of expertise, and the restrictions of your students.

Though research shows that most of your students will be self-motivated, there is a definite preference for some kind of interaction, both with the faculty member and other students. You might plan to have either a review session, group discussions, or occasional seminars.

Pre-course preparation is crucial. There are a series of questions it is important for you to consider before planning your course.
* What are the course objectives?
* What are your academic expectations of the students? What kind of performance do you expect?
* How will you communicate these objectives and expectations to the students?
* What administrative procedures will work best for your students and yourself?

Know the Course:
In addition to reading and being familiar with this *Faculty Guide*, you should be familiar with the other components of this course. Understand how the course components have been carefully integrated. Watch the videos to acquaint yourself with the various topics presented. Familiarize yourself with the selected textbook (*Microbiology*, 4th ed., Prescott/Harley/Klein, McGraw-Hill) and the *Telecourse Student Guide* to accompany it.

Fit the Course into Your Curriculum:
Know the course number, the prerequisites, the programs, and the credits that the students earn upon successful completion of the course.

Understand the Procedure:
Inform yourself about the broadcast schedule, registration, orientation, expected enrollments, discussion meetings, testing schedules, and student support services. Find out who is to make the above arrangements and how they are to be communicated to the students. It may be you! Make sure you know your responsibilities and to whom to go to for assistance if problems occur.

Adapt this Course to Your Needs:
Review the key points, lesson overview, student assignments, and web exercises in the guide to decide what to assign as required assignments and what to assign for extra credit. You may want your students to reflect on some of the Critical Thinking problems in the textbook.

Read the questions provided. Adapt them as you see fit, or use the complete test bank as is. Check the text web site for additional instructor material, quizzes, and web exercises.

Keeping Up

Just as in any other college course, you are a key factor for student success. Many students will look to you for help, many others will drop out if you do not reach out to them. Keeping up with the course and the students is crucial.

Keeping Up With the Course:
Watch each video and follow each telecourse student guide chapter, so that you will be able to answer questions and discuss each topic with the students who need extra help. Revise the tests as needed to fit your students. Supervise the assignments for which you offer extra credit.

Keep Up With the Students:
Keeping good contacts with your students is key to the success of the telecourse. Maintain contact with students through newsletters, e-mail reports on progress, and personal or telephone conferences. Develop an informal style of correspondence with students to provide a sense of personal interest in their progress.

Conduct study sessions and discussion meetings as you think necessary; before and after tests are a good time for those meetings. Write out your suggestions and criticisms fully on each assignment. Develop other activities as you see the need. Keep in mind that requiring many on-campus activities may defeat the purpose of this delivery system. Other student communication devices are usually more effective.

Recording and Reporting:
Although every teacher has to be concerned about keeping accurate records, this part of a telecourse instructor's job is especially important and time consuming. It is also quite possible that you will never see some of your students face-to-face. Your contact may all be via telephone, e-mail, and paper – perhaps only on paper.

Consequently, it is especially important that you keep full and complete records of all student contacts, activities, deficiencies, and successes, so that you can help students appropriately and evaluate them fairly. Keep track of students who attend orientations and discussion meetings, call on the telephone, send you e-mail messages, or come by for conferences.

Be very particular about recording test scores and other grades that figure into the final evaluations of students. Be prompt in returning written assignments and in sending out grades after each test. Supervise grade records closely if you do have clerical assistance.

Help Your Students Get a Good Start:
Telecourse students need special help from you. Since they will not regularly have the benefit of interaction with you and their classmates in a group, they will rely heavily on you to inform them about procedures and requirements. At the same time, they may be reluctant to ask questions – they may not know the right questions to ask. You can help by reaching out to them.

Supply an Information Bulletin or Booklet:
Make sure your students receive, at registration or before, a viewing schedule for the videos and information on the other ways the telecourse programs for the course will be available to them.

Include the times, dates, and places for discussion groups and examinations, general information about the course; a description of student services available to them; a list of required books; and grading policies. Communicate this information to them via e-mail as well as in a printed page form.

Orientation Letter:
Welcome your students to the telecourse, reassure them that they can both succeed and find this course stimulating, challenging, and rewarding. Use this letter to express your own interest and enthusiasm in the course.

Give students your office phone number, e-mail address, your office address and location (with directions) and the times you will be available. It is also helpful to provide the students with some background information about yourself, your field of knowledge, experience, and interest in this particular microbiology course.

Remind them about the broadcast schedule and orientation sessions and stress the importance of attending. Tell them to acquire your institution's *Telecourse Information Booklet* (see the section on preparing a *Telecourse Information Booklet*) if they do not have one already. Identify the selected text and study guide; *Microbiology, 4e* by Prescott/Harley/ Klein and *Telecourse Student Study Guide t/a Unseen Life on Earth: An Introduction to Microbiology,* by John P. Harley. Alert them to the location and operating hours of the bookstore. Mail the orientation letter as soon as possible or hand it out at registration.

Orient Students to the Telecourse:
Inform your students about the advantages and disadvantages of taking a distance-learning course and about the best ways to approach telecourses. Identify whatever support services your institution offers to telecourse students. You may want to give them this information at orientation, in a newsletter, or via e-mail (or in all of the preceding – a little reinforcement never hurts).

Provide a Student Syllabus:
The student syllabus will provide students with the specifics about requirements and responsibilities for the telecourse. If this is the only telecourse offered, the syllabus and information booklet may be combined, but if several courses are included in the general information piece, students will feel more comfortable with a special syllabus explaining this course.

Plan Ahead for a Good Orientation:
It is the responsibility of the campus instructor to plan a good orientation meeting. You may have to schedule rooms, dates, and times, or you may have to prepare only the content. Plan to hold the orientation before the course is locally broadcast, if possible. Remember that telecourse students will probably be able to attend evening and weekend meetings more readily than weekday or daytime ones. Make sure that the students are informed about this meeting (if possible, provide a campus map or directions to the meeting). Research shows a high correlation between attendance at orientation and successful completion of the telecourse. Make the orientation mandatory and make it worth attending.

Scheduling the Orientation Meeting:
Ideally, the orientation meeting will fall between the last day of registration and the first broadcast date. Since many telecourse students work, plan these sessions on the weekend and at several times during the day, if possible. For an expected enrollment of 200, you will need at least two meetings. More would not be out of line.

Be sure the meeting room is an adequate size and that it is well marked. Signs in the lobby and hallways help new students feel more comfortable that they are headed in the right direction. Start promptly and finish on time. Allow at least one hour for this meeting.

Establish the Student Roll:
It is a good idea to ask students to fill out index cards with their names, social security numbers, addresses, e-mail addresses, phone numbers, best time to reach them, and other information you would like to have. Easily alphabetized and stored, these cards will prove very helpful all through the course. Think of ways the students can be comfortable with each other and encourage them to exchange phone numbers or e-mail addresses to create a learning support system.

Introduce Yourself:
Be sure to introduce yourself and establish the kind of role you see for yourself throughout this course. Express your interest and enthusiasm about the course topic and structure.

Tell the students your office number and location, your e-mail address, your campus phone number, your office hours, and the way they can leave messages for you when you are not in your office.

Introduce the Course:
Even though the student should have received an information bulletin with their registration materials, make sure you have plenty of copies to distribute now. Include information about the course, broadcast schedule, test and discussion meeting schedules, necessary books, grading practices, and other information you think is helpful. Tell them something about the course – the purpose, the objective, the outline, and some highlights.

Find Out What the Student Know About this Subject:
It may be helpful to know what background, if any, the students have in the basic sciences. You may want to know why they are taking microbiology, and what their career goals are going to be. Selecting examples for problems or topics for special projects that reflect their interests will help them succeed in the course.

Describe Distance-Learning Instruction:
Orient the students about distance-learning instruction. First, the distinct advantages: convenience, economy in time and travel, opportunities for making up missed sessions, and evening lessons. Then discuss the disadvantages – the extra self-discipline required, the lack of immediate answers to questions, the reduction in student-to-teacher and student-to-student contact.

Dispel the myth that taking a distance-learning course is easier than taking a course on campus. Encourage the students to establish a viewing schedule and the proper viewing environment –

slouching around, drinking a soda while chatting with family is not the best way to absorb a lecture in their own home. Advise them to listen and watch intently and to rely on their memories, not their notes. The pace and visualization of television make taking more than the briefest notes unsatisfactory, if not impossible.

Explain the Course Components:
Explain that taking the telecourse requires more than just watching a series of television programs. Explain each course element:
- Television programs
- Assigned readings from *Microbiology,* 4th ed., Prescott/Harley/Klein
- *Telecourse Student Guide* to *Unseen Life on Earth: An Introductionto Microbiology* chapters
- Discussion meetings
- Lab sessions (if offered)
- Tests
- Newsletters, e-mails
- Extra credit assignments
- Campus instructor
- Hotline

Recommended Skills for Taking an Online Course
If you are interested in registering for an online course, you are expected to have the following minimal access skills:

- Access to a computer with at least Windows 95 (or equivalent), a 28.8 modem, and 16 MG or more of RAM (or a Power Mac)
    - (1) a telnet client such as Pueblo
    - (2) an FTP client
- A high school, college, or university VAX account or a personal account with a commercial carrier
- Ability to word-process and manage text files
- Ability to send, receive, reply to, forward, and save e-mail; e.g., using PINE or NETSCAPE Mail
- Reliable access to the Internet via an Internet Service Provider (ISP)
- Ability to use NETSCAPE or MICROSOFT EXPLORER as a graphical browser in order to
    - (1) do searches
    - (2) bookmark and organize Web sites
    - (3) download programs
    - (4) copy text and graphics from a Web site
- Ability to cut and copy text from one application to another

If you are unable to do one or more of the above, you should gain such skills before registering for an online course.

It is also desirable to have these additional skills:
(1) Ability to use FTP
(2) Ability to compose HTML documents with an HTML editor such as Adobe PageMill, NETSCAPE NAVIGATOR GOLD (or COMMUNICATOR)

Newsletters, E-mail:
Periodic newsletters, available on paper or via e-mail, are very useful ways of keeping in touch with students. Ideally, one newsletter will be sent as soon after registration as possible. Reinforce what you cover in orientation. A newsletter before each examination is also very helpful for the students, stressing the key points for study.

Inform the students about on-campus activities and services that they might need or enjoy and about cultural events in the community that might be on interest to them.

Plan for Discussion Meetings:
Discussion meetings before and after tests are very helpful for students, particularly for those who miss the traditional classroom exchange with other students. Be sure that students are informed about the meetings and that the meetings are worth attending.

It is not essential that students be required to attend the discussion meeting. While there is a high correlation between attendance at orientation and successful completion of the course, there is no such evidence for attendance at discussion meetings. However, many students like them and find them helpful.

Suggestions for Content:
Discussion meetings should focus on areas of students' interest and needs for taking *Unseen Life on Earth An Introduction to Microbiology* telecourse:
Questions about course content
- Discussion of career goals and the applications for topics they have studied
- Guest speakers from area labs and industries, or public health agencies
- Lab session to reinforce material presented in the videos and text
- Discussion of upcoming test
- Discussion of last test results
- Questions about any critical thinking assignments you have given

As the campus instructor, use your imagination and plan whatever you think will be important and beneficial for your students

Guidelines for Administrators:

Administrative Planning:
While the production quality and instructional effectiveness of a telecourse are of paramount importance, efficient course administration is at least equally important. Clear-cut and carefully followed procedures are necessary for the telecourse to work smoothly. Good planning, well in advance of the first broadcast, by a team of administrators, faculty, and support personnel, is the key to a good telecourse operation.

Ideally, everyone associated with the course should read the entire *Faculty Manual*; the team should then decide on the ways that the following suggestions can be best adapted to the needs of your college.

Overview of Administrative Concerns:
The administration of any telecourse must take into account all of the following functions. Many of these responsibilities will be delegated to other college operations: but, all must be considered in the overall management system for this telecourse.

1. Arrange and schedule air time
2. Promotion (planning, budget, creation of ads, press releases, brochures, other promotional materials, procedures for handling information requests).
3. Ordering materials (videotapes, *Telecourse Student Guide*, text books)
4. *Telecourse Information Booklet* (collection of information, composition, duplication, distribution to students and college staff).
5. Staffing (telecourse instructor(s), supervisor, support personnel, supplies, office space, procedures).
6. Counseling of students (procedures, personnel, times, and place).
7. Course content (use of Student Study Guide, text, related readings, lab sessions, assignments).
8. Registration of students (procedures, supplies, personnel, and place).
9. Class rolls (division of students, if necessary, preparation or rolls, certifying of rolls, preparation of address lists for mailings).
10. Orientation of students (procedures, supplies, personnel, times, and place).
11. Telephone communication (hotline service, procedures, supplies, personnel, times, and place).
12. Newsletters (preparation, duplication, distributions, e-mail carriage).
13. Course Web Site (preparation, server space for posting, supervision and maintenance, procedures, personnel).
14. Discussion sessions for students (procedures, supplies, personnel, times, and place).
15. Lab sessions (planning, supplies, procedures, personnel, times, and place).
16. Testing (procedures, supplies, personnel, times, and place).
17. Procedure for dealing with student problems (personnel, policy, time, and place).
18. Fulfilling contractual responsibilities (submission of enrollment figures, payment of fees, provision of numbers, kind, and location of video tapes, erasure of tapes at termination of contract, renewal of contract).
19. Evaluation of course (approval and distribution of evaluation instruments, collection and analysis of data).
20. Alternate uses of telecourses

Broadcast Time:
Be sure that you or the appropriate person at your institution have made proper arrangements with your local public television broadcasting station and the PBS Adult Learning Service to use *Unseen Life on Earth*: *An Introduction to Microbiology* telecourse. Find out the exact information about the day and time for local broadcast.

Promotion of the Course

Importance of Promotion:
Good promotion is more critical for gaining enrollment in telecourses than it is in gaining enrollment for on-campus courses. Thus, it is important for the administrator to plan for the personnel, time, and money necessary for an effective promotional effort.

Promotion Ideas:
- Consider these promotional ideas:
- Press releases about the course, instructor, college involvement, community tie-ins with the course, and the delivery method (television).
- Public service announcements on television and radio.
- Announcements on your college Website
- Brochures and flyers announcing this course or your entire telecourse operation – mailed, placed in public libraries, supermarkets, and theaters.
- Mass mailing to residents of selected zip codes or to targeted audiences such as employees of local hospitals, or teachers groups.
- Posters
- Letters to personnel directors of large companies requesting a mention of the course in the company newsletter.
- Newspaper ads, especially on the television page.
- Ask your public television station to help you promote your course.
- Link your promotion to other college program advertising campaigns.
- Offer a unit from your course as the program for a community organization meeting.

In-house Promotion:
While the most expensive promotion for your telecourse operation will be related to informing the community, informing your faculty and staff is equally important.   Everyone in the institution from the clerical staff to the highest administrator needs to know that telecourses in general and this telecourse in particular have been added to your program. Those who answer the phone, those who advise students before they register, and those who conduct registration need to know all of the particulars about your telecourse program so that they can efficiently help students.

The Best Promotion:
The very best promoter of a telecourse is a satisfied student. A smoothly operating course which gives students satisfaction while they take it and affords them success at its conclusion is the best advertisement your college can have. The reverse is also true. Disgruntled students complaining about their dissatisfaction can damage a telecourse program.

Telecourse Information Booklet

Purpose:
A *Telecourse Information Booklet*, which serves as a general guide to your telecourse(s), is a vital tool for informing and enrolling students. Telecourses are, after all, a different way to learn and earn college credit. While serving as an informational and motivational tool for prospective telecourse students, the booklet also gives students important information about registration, tests, texts, seminars, labs, and other activities.

Content:
1. A *Telecourse Information Booklet* should contain information about the following items:
2. Distance-learning as it applies to your institution.
3. All distance-learning offerings complete with the course descriptions (name; official number; credit hours; prerequisites, if any; transferability; and general goal or area of concern).

4. Registration procedures
5. Broadcast schedules and information about other ways in which television programs may be available (videos that can be checked out of the library, etc.).
6. Tuition and fees, including refund policy.
7. Schedule and explanation of events relevant to each course – orientations, discussions meetings, labs, tests, make-up test policy, etc.
8. Hotline phone service
9. Text and study guides and other materials needed for each course
10. Newsletters, e-mail
11. Student services (library, counseling, study skills lab, health services, etc).
12. Evaluation procedures
13. Address information (importance of keeping address current; procedure for changing address records).

Staffing

Introduction:

Staffing is of primary importance for the success of a telecourse. Depending upon the size of your telecourse operation, a very few to a great many staff members may be needed. All of the following positions are considered useful in an effective telecourse operation. However, the division of labor represents a suggested plan, not a mandate. Certainly, some of these duties can be combined in an operation with a relatively small enrollment.

Telecourse Instructor:

Whether the telecourse is a part of the assignment for a full-time instructor or whether you hire someone especially for the position is not as important as choosing a teacher who is enthusiastic about the use of television. This instructor should have the principle role in the management of this course. The instructor is responsible for assisting students enrolled in the telecourse, for maintaining the students' records and grades, and for contribution to the local adaptation and improvement of the telecourse. The instructor also serves as a resource person for the rest of the faculty and the community about this telecourse (see example position description for Telecourse Instructor in this section).

Clerical Assistance:

It is always helpful and usually cost-effective to provide clerical assistance for the telecourse instructor. One instructor can effectively handle a much larger number of students in a telecourse than in most on-campus classes is there is adequate support service. Assistance with word processing, duplicating, collating, distributing, filing, and recording will save the instructor's time for working with students.

Coordinating Instructor:

When enrollments require more than one instructor for *The Unseen Life on Earth* telecourse, or when more than one campus in a district or a consortium is involved, it is helpful to designate a coordinator of the course. (See sample position description for Coordinating Instructor in this section).

Telecourse Counselor:

One counselor on the regular college staff should be designated as the telecourse student counselor. The counselor should be responsible for advising potential telecourse students, assisting in individual student program planning, assisting in diagnostic testing of students, and counseling enrolled students as requested by student or instructor referral. The counselor should serve as the advocate for the telecourse students and as the resource person to inform other counselors about telecourses in the counseling staff meetings. Telecourse students need not be restricted to seeing this one counselor, but one counselor should be particularly informed and available.

Test Center Personnel:
If large numbers of telecourse students are enrolled, a testing center is probably required. If your college already has a test center, its operation can perhaps be expanded to include telecourse students. If you do not have such a center, one may be established in any suitable room as needed for telecourse students. Paraprofesisonals make excellent proctors and supervisors of testing procedures. It works well to have two people on hand at all times if numbers are large and testing hours are extensive.

Telephone or E-mail Hotline:
It is very desirable to provide a telephone or e-mail information service for telecourse students. In a small operation, the instructor may cover the e-mail messages and telephone. However, if there are large enough enrollments, a separate telephone staffed from morning through night during school hours, is advisable. The operator(s) should be trained to answer a variety of questions about current and upcoming telecourses. Subject matter questions should be referred to campus instructors. For your operator(s), select someone who has a pleasant telephone personality and answers questions in a helpful, considerate manner.

Coordinating Instructor: Position Description

The Coordinating Instructor is the person employed to coordinate course activities for two or more colleges or campuses or two or more instructors in the district or consortium. The coordinator's responsibilities include the following:
1. Thorough familiarization with every aspect of course content, operation, and administration
2. Assignment of tasks to appropriate personnel as needed to insure smooth operation of the course.
3. Management of the various aspects of the course administration, including instructor's responsibilities, and those of paraprofessionals, student assistants, clerical help, and others that may be assigned to the course.
4. Interviewing, recommending for employment, supervising, and evaluating instructors, paraprofessionals, and assistants.
5. Coordinating and supervising the course's portion of telecourse registration, late registration, and mail-in registration.
6. Adjusting class section size after class rolls are validated, as needed.
7. Validating class rolls and maintaining student participation records for audit purposes.
8. Coordinating orientation sessions and discussion meetings.
9. Ensuring that a testing center is adequately prepared with necessary materials and procedures prior to each testing period.
10. Ensuring that such lab space as need is adequately prepared, with equipment and supplies, at the start of each term

11. Coordinating procedures for make-up tests, withdrawls, and other student activities.
12. Meeting periodically with the Dean of Instruction and Division Chairpersons to solve problems related to efficient management of telecourse-related activities.
13. Coordinating the promotion schedule for television courses.

Telecourse Instructor: Position Description

The campus Telecourse Instructor's job is to assist students enrolled in the telecourse at the college, maintain student records and grades, and contribute to the success and improvement of telecourse. Specific responsibilities include the following:

1. Writing and distributing pre-enrollment and orientation information as well as placing it on the college's Website. This includes vital information on the operation of the course and explicit explanation of what is expected. The information should be available before the orientation and may be presented in the form of a *Telecourse Information Booklet* (see section on preparing a *Telecourse Information Booklet*).
2. Writing, mailing, or e-mailing a welcome letter to students.
3. Knowing all elements of the course and how they have been integrated.
4. Maintaining posted office hours during which students may make personal contact by phone or in person, or during which the instructor may return student calls and e-mail messages.
5. Revising and modifying tests as necessary.
6. Writing periodic newsletters and distributing them to the students on paper and via e-mail.
7. When more than one instructor is involved in the district or on campus, newsletter preparation responsibility should be shared.
8. Grading examinations and other assignments as appropriate: maintaining grade records on each exam and assignment.
9. Preparing for and leading orientation sessions for students enrolled in the course.
10. Maintaining student records.
11. Making decisions and arrangements regarding make-up tests, withdrawls, and other problems students may encounter. Such decisions must be in accord with college rules and policies.
12. Meeting periodically with the Coordinating Instructor and other campus instructors in the course for the purpose of evaluating the elements of the course and recommending changes.
13. Conducting student evaluations of the course as requested by supervisors.
14. Supervising clerical assistant.
15. Setting up and supervising on-campus exams.
16. Ordering and maintaining sufficient quantities of on-campus materials to keep course running smoothly.

Registration

Introduction:

Smooth registration for a telecourse is as important as it is to other college programs. A well-informed faculty and counseling staff and well-planned procedures are a must.

Mail-in Registration:

Even if your college does not normally use mail-in registration, it might be useful to investigate the procedure for telecourses. One of the purposes of telecourses is to save people the drive to and from campus, and registration by mail supports that convenience.

One good way to handle registration by mail is to send students, upon request, the application and registration forms with instructions to send these forms and the appropriate fees to the admissions office of the college. The registrar can then record the enrollment and mail the student a receipt and the special materials and information a telecourse student requires.

Website Registration:
Another possible method to provide the student with the convenience of off-site registration is to allow them to request registration information on telecourses from the college's website. Consider allowing them to actually complete a pre-registration form on-line, which is then finalized with the receipt of their fees.

On-campus Registration:
Most telecourse students will enroll on campus during the regular registration period of the college. Professional personnel, knowledgeable about distance-learning courses, should be available during registration to advise and counsel student about taking a course via television.

If possible, a separate registration table should be set up, clearly labeled, and looked after by someone familiar with the particular course and with instructional television. Viewing times should be displayed and informational material should be distributed to all who enroll.

If a special telecourse registration table is not available, all of those who process registration should be oriented about the nature and requirements of telecourses so that they can answer students' questions and give good advice.

Registration Materials:
Student should be handed the following information about the telecourse if they register on campus or mailed this information if they register by mail:
- Course requirements
- Viewing schedule
- Dates, times, and places for exams, discussion meetings, labs, orientations, and other pertinent events.
- Name, phone number, e-mail address and office hours for instructor for the course.
- List of required text and study guide for *The Unseen Life on Earth*. (include mail order information if the college store will accept such orders. If mail order purchases are not possible, inform the student of the days and times the campus book store is open).
- College services available to students – counseling, library, media, study skills center, and other services of value to distance learners.

Roll Cards:
At registration students should fill out a roll card giving their names, addresses, telephone numbers, e-mail addresses, and social security or student identification number. If students register by mail, the person accepting the registration should fill out such cards and forward them to the instructor. Such cards should form part of the enrollment packet.

Existing Campus Services
Telecourse students should be entitled to the same college services as those afforded to on-campus

students. Such students will need special information about these services, and the on-campus personnel will need to be aware of the distance-learning students.

If these students are to take advantage of what the college has to offer, you and your staff will have to make a special effort to inform them about just what is available in terms of counseling, study skill labs, library facilities, entertainment, health services, and recreation.

Encourage all campus services to include telecourse students in their promotions and planning. Encourage the telecourse instructors to inform the students about the other facilities of the institution in their newsletters and discussion meeting(s).

PART II: GUIDE FOR INDIVIDUAL PROGRAMS

Microbial Cell Biology
Program 1
The Microbial Universe

It was the knowledge of the great abundance and
wide distribution of actinomycetes, which dated back nearly three
decades, and the recognition of the marked activity of this
group of organisms against other organisms that led me in
1939 to undertake a systematic study of their ability to
produce antibiotics.

–Selman A. Waksman

One of microbiology's fifty most significant events in the past 125 years:

1944

Albert Schatz, E. Bugie, and **Selman Waksman** discovered streptomycin, soon to be used against tuberculosis. Streptomycin has the same specific antibiotic effect against gram-negative microorganisms as penicillin does on gram positives. **Waksman** was awarded the Nobel Prize in Physiology or Medicine in 1952.

Student Assignments

- **Watch** Video 1: The Microbial Universe
- **Read** Chapter 1: The History and Scope of Microbiology , and
    Chapter 2: The Study of Microbial Structure: Microscopy and Specimen Preparation
        in *Microbiology*, 4th ed, Prescott/Harley/Klein, McGraw-Hill
- **Lab Experiments** from *Laboratory Exercises in Microbiology*, 4th ed, Harley/Prescott, McGraw-Hill
    Exercise 1: Bright-Field Light Microscope
    Exercise 2: Dark-Filed Light Microscope
    Exercise 3: Phase-Contrast Light Microscope
    Exercise 4: Fluorescence Microscope
    Exercise 5: Microscopic Measurement of Microorganisms

16

Lesson Overview

Video 1: The Microbial Universe
The world of microorganisms is a dynamic one, and all other life forms depend on microbial metabolic activity. Recent genetic studies lead the student to one of approximately 99 percent of "undiscovered" microorganisms.

Chapter 1: The History and Scope of Microbiology
This chapter introduces the field of microbiology and discusses the importance of microorganisms not only as causative agents of disease but also as important contributors to food production, antibiotic manufacture, vaccine development, and environmental management. It presents a brief history of the science of microbiology, an overview of the microbial world, and a discussion of the scope and relevance of microbiology in todays society.

Chapter 2: The Study of Microbial Structure: Microscopy and Specimen Preparation
This chapter provides a relatively detailed description and operation principles of the bright-field microscope. Other common types of light microscopes are also described. Following this the students are introduced to various procedures for the preparation and staining of specimens in order to observe general features and/or specific structures. The chapter concludes with a description of the two major types of electron microscopes and the procedures associated with their use.

Learning Objectives

After successfully completing this unit, the student will be able to:

- define the science of microbiology and describe some of the general methods used in the study of microorganisms
- discuss the historical concept of spontaneous generation and the experiments that were performed to disprove this erroneous idea
- discuss Koch′s postulates, which are used to establish the causal link between a suspected microorganism and a disease
- describe some of the various nonpathological activities of microorganisms
- discuss the importance of the field of microbiology to other areas of biology and to general human welfare
- describe how lenses bend light rays to produce enlarged images of small objects
- describe the various parts of the light microscope and how each part contributes to the functioning of the microscope
- compare the images obtained by the different types of light microscope
- describe the preparation and simple staining of specimens for observation with the light microscope
- describe the Gram-staining procedure and how it is used to categorize bacteria

- describe the basis for the various staining procedures used to visualize specific structures associated with microorganisms
- compare the operation of the transmission and scanning electron microscopes with each other and with light microscopes

Course Integrator

Chapter 1
The Discovery of Microorganisms (p. 2)

Web Link	Antony van Leeuwenhoek: http://www.ucmp.berkeley.edu/history/leeuwenhoek.html
Web Link	*Microbiology* Online Learning Center: http://www.mhhe.com/biosci/cellmicro/prescott/

The Spontaneous Generation Conflict (p.2)

The Recognition of the Microbial Role in Disease (p.7)

The Discovery of Microbial Effects on Organic and Inorganic Matter (p.9)

Microbiology in the Twentieth Century (p. 11)

Web Link	Centers for Disease Control: http://www.cdc.gov/
Web Link	Emerging Infectious Diseases: http://www.cdc.gov/ncidod/EID/index.htm
Web Link	World Health Organization: http://www.who.int/
Web Link	Bugs in the News: http://falcon.cc.ukans.edu/~jbrown/bugs.html

The Composition of the Microbial World (p. 13)

Web Link	The Curious Microbe: /dlc-me/curious/cindex.html
Web Link	The Microbial Zoo: http://commtechlab.msu.edu/sites/dlc-me/zoo/
Video *Unseen Life on Earth*	Unit #1 The Microbial Universe
Video *Unseen Life on Earth*	Unit #7 Microbial Diversity

The Scope and Relevance of Microbiology (p.13)

Chapter 2
Lenses and the Bending of Light (p.17)

Web Link	*Microbiology* Online Learning Center: http://www.mhhe.com/biosci/cellmicro/prescott/

The Light Microscope (p. 18)

Visual Resource Library	Bright-Field Microscope Figure 2.3
Visual Resource Library	Microscope's Light Path Figure 2.4

Visual Resource Library	Phase-Contrast Microscopy Figure 2.9
Visual Resource Library	Contrast in Phase Microscopy Figure 2.10
Visual Resource Library	Fluorescence Microscopy Figure 2.11
Visual Resource Library	Microscopic Resolution Limits Figure 2.19
Hyperclinic	Diagnostics: Direct Examination
Web Link	Micscape Online Microscopy Magazine: http://www.microscopy-uk.org.uk/mag/indexmag.html

Preparation and Staining of Specimens (p. 24)

Visual Resource Library	Gram-Staining Procedure Figure 2.13
Visual Resource Library	Microscopic Resolution Limits Figure 2.19
Hyperclinic	Diagnostics: Direct Detection
Microbes in Motion II	Bacterial Structure & Function Book:
Bacteria Groups/Introduction	
Bacteria Groups/Categories	
Bacteria Groups/Classification	
Web Link	Gram Stain: http://www.meddean.luc.edu/lumen/DeptWebs /microbio/med/gram/gram-stn.htm

Electron Microscopy (p. 27)

Visual Resource Library	Scanning Electron Microscope Figure 2.26
Web Link	How the SEM works: http://www.mos.org/sln/sem/

Microbial Terms Used in Video 1 in Order of Appearance with Page Indexing to Textbook (*Microbiology*, 4th ed, Prescott/Harley/Klein, McGraw-Hill)

biomass 841	herbivore 566	seaweed 548
ecosystems 834	pathogens 582	carbon cycle 837
microbes 571	disease 721	primary producers 832
microscopes 18	disinfectant 137	diatoms 547
protozoa 544	vaccines 607	Sargasso Sea 542
algae 541	tuberculosis 773	nucleotides 203
fungi 396	Alexander Fleming 678-79	plankton 541
bacteria 39	pneumonia 802	food web 554
viruses 342	syphilis 786	spread plate 107
electron microscope 27	Hanta virus 728	colony 107
nitrogen fixation 198	photosynthesis 182	
	cyanobacteria 443	
	recycling 886	

Web Links Related to Video 1 (The Microbial Universe) and Readings in Chapter 1 (The History and Scope of Microbiology) (Go to www.mhhe.com/prescott, Student on Line learning Center, Double Click on Chapter 1, Double Click on Web Site)

Anthony van Leeuwenhoek (1632-1723)
http://www.ucmp.berkeley.edu/history/leeuwenhoek.html

The Curious Microbe. An essay of the extreme and the unusual.
http://commtechlab.msu.edu/sites/dlc-me/curious/cindex.html

Microbes in the News. News stories from the popular press related to microbial ecology and microbiology, with added commentary from scientists. Stories are grouped into four categories: heroic, dangerous, ancient, and strange.
http://commtechlab.msu.edu/sites/dlc-me/news/news.html

Microbiology, Bugs in the News. Read about cryptosporidium, AIDS, HIV, flu, *E. coli*, the immune system and lots more.
http://falcon.cc.ukans.edu/~jbrown/bugs.html

Dictionaries and Glossaries (American Society for Microbiology). Links to the biomedical and life science resources that provide the spelling of proper names and definitions.
http://www.hardlink.com/~tsute/glossary/index.html

Microbiology. Weekly microbiology articles and links.
http://www.suite101.com/welcome.cfm/microbiology

Web Links Related to Video 1 (The Microbial Universe) and Readings in Chapter 2 (The Study of Microbial Structure: Microscopy and Specimen Preparation) (Go to www.mhhe.com/prescott, Student on Line Learning Center, Double Click on Chapter 2, Double Click on Web Site)

Microscape–the Online Microscopy Magazine. This web site includes images, general articles, popular articles, and microscope reviews. Highly recommended.
http://www.microscopy-uk.org.uk/mag/indexmag.html

Gram Stain. A teaching page on the gram stain where you can view gram-stained images of bacteria that are medically important.
http://www.meddean.luc.edu/lumen/DeptWebs/microbio/med/gram/gram-stn.htm

Scanning Electron Microscope. Museum of Science, Boston. How the SEM works and an image gallery.
http://www.mos.org/sln/sem/

Dennis Kunkel's Microscope. Science and photography through a microscope. Spectacular electron micrographs.
http://www.pbrc.hawaii.edu/~kunkel/index.shtml

Microbial Terms Used in Chapter 1 (The History and Scope of Microbiology)

eucaryotic cell   13
hypothesis 8
Koch's postulates 8

microbiology 2
microorganism 2
procaryotic cell   13

spontaneous generation   2
theory 8

Microbial Terms Used in Chapter 2 (The Study of Microbial Structure: Microscopy and Specimen Preparation)

acid dyes   25
fluorescent light 23
phase-contrast
microscope 22
acid-fast staining   25
atomic force
microscope 33
basic dyes   25
bright-field
microscope 18
chromophore
groups   25
dark-field
microscope 21
dark-phase contrast
microscopy 22
differential staining
procedures 25
eyepieces 18
fine and course
adjustment knobs 18
fixation 25
flagella staining 27
fluorescence
microscope 23

fluorescent light   23
fluorochromes   24
focal length   17
focal point   17
freeze-etching 31
Gram stain 25
immersion oil 19
mordant 25
negative staining 25
nosepiece 18
numerical aperature 19
objectives 18
oculars 18
parfocal 19
phase-contrast microscope 22
refraction 17
refractive index   17
resolution 19
scanning electron
microscope 31
scanning probe microscope 33

scanning tunneling
microscope 33
shadowing 30
simple staining 25
spore stain 27
stage 18
substage condenser 18
transmission electron
microscope (TEM) 28
working distance 20

Sample Test Questions from Video 1 (The Microbial Universe)

1. The organic matter produced by living organisms and measured by weight is termed __________.

2. An ecological unit consisting of both the biotic (living) and nonliving environment which interact to produce a stable system is an __________.

3. The human body contains approximately _____ trillion microbes.
   a. 1
   b. 10
   c. 100
   d. 1,000

4. Which of the following is not a living cell?
   a. protozoa
   b. algae
   c. fungi
   d. bacteria
   e. viruses

5. In order to see the bacteria on the head of a pin, you need to use a
   a. dissecting microscope
   b. compound microscope
   c. electron microscope
   d. scanning electron microscope
   e. Both c and d

6. Herbivores, such as a cow, do not need microbes in order to break down the cellulose in the grass they eat.
   a. True
   b. False

7. Those microbes that cause disease are called:
   a. infectious
   b. pathogens
   c. disease producers
   d. Both a and b

8. In 1928, Alexander Flemming discovered that __________ was producing a diffusible substance lethal to staphylococci.
   a. *Streptomyces*
   b. *Penicillium*
   c. *Cephalosporium*
   d. *Bacillus*

9. Which of the following is the most recent disease to be discovered?
   a . TB
   b. pneumonia
   c. Hanta virus disease
   d. syphilis

10. Which of the following were the first microbes to carry out photosynthesis?
    a. cyanobacteria
    b. bacteria
    c. fungi
    d. algae
    e. green plants

11. For as long as human beings have been around, we and microbes have been evolving together.
    a. True
    b. False

12. Without microbes, life itself on the planet earth would be as barren as the surface of the moon.
    a. True
    b. False

13. Since microbes have been around for such a long period of time, they do not shuffle their genetic information.
    a. True
    b. False

14. In DNA, the instructions for all the wonderful and complex organisms every produced, including we humans, contain just _____ different chemicals.
    a. two
    b. three
    d. four
    e. five

15. In DNA, the four basic chemicals have been arranged into an almost infinite number of combinations to produce what scientists call a ___________.

Sample test Questions from Text Chapter 2 (The Study of Microbial Structure: Microscopy and Specimen Preparation) Web Site

1. In a compound microscope, the primary image is formed by an objective lens and enlarged by the optical lens to reveal a virtual image.
   O True
   O False

2. Microscope resolution increases as the wavelength of radiation used to illuminate the specimen decreases.
   O True
   O False

3. The substage condenser
   O changes the wavelength of the light reaching the specimen
   O focuses light on the specimen
   O decreases the amount of light reaching the specimen

4. A 40X objective and an 10X ocular produce a total magnification of
   - O 50
   - O 400
   - O 90

5. Living, unstained cells and organisms can be observed best using
   - O fluorescent microscopy
   - O TEM
   - O phase contrast microscopy

6. Scanning electron microscopy (SEM) is best used to study
   - O small internal cell structures.
   - O surface morphology.
   - O all of the above

7. The fluorescence microscope has become an essential tool in medical microbiology to identify bacterial pathogens.
   - O True
   - O False

8. A microscope in which an image is formed by passing an electron beam through a specimen and focusing the scattered electrons with magnetic lenses is called a
   - O transmission electron microscope
   - O scanning electron microscope
   - O phase-contrast microscope

9. Resolution is the ability of a lens to distinguish between small objects close together. What approximate resolution can be obtained with a lower power (10X) objective lens?
   - O 0.2 microns
   - O 0.3 microns
   - O 0.9 microns

10. Fixation is the process by which the structures of the cells are preserved and fixed in position. An advantage of chemical fixation over heat fixation is that it
   - O magnifies the specimen
   - O does not destroy internal structures

Chapter 1 Sample Test Questions on The History and Scope of Microbiology

1. Which of the following is (are) used to define the field of microbiology?
   a. the size of the organism studied
   b. the techniques in the study of organisms regardless of their size
   c. both the size of the organism studied and the techniques employed in the study of organisms regardless of their size
   d. neither the size of the organism studied nor the techniques employed in the study of organisms regardless of their size

2. Who of the following developed a set of criteria that could be used to establish a causative link between a particular microorganism and a particular disease?
   a. Fracastoro
   b. Koch
   c. Pasteur
   d. Lister

3. Who of the following was the first to observe and accurately describe microorganisms?
   a. Pasteur
   b. Lister
   c. van Leeuwenhoek
   d. Tyndall

4. Who of the following provided the evidence needed to discredit the concept of spontaneous generation?
   a. Pasteur
   b. Koch
   c. Semmelweiss
   d. Lister

5. Who of the following provided strong evidence that DNA was the genetic material and carried the genetic information during transformation?
   a. Luria and Delbruck
   b. Avery, MacLeod, and McCarty
   c. Beadle and Tatum
   d. Hershey and Chase

6. The concept that living organisms arise from nonliving material is called
   a. biogenesis.
   b. cell theory.
   c. spontaneous generation.
   d. relativity.

7. Cells with a relatively simple cell morphology that do not have a true membrane-delimited nucleus are called
   a. procaryotes.
   b. eucaryotes.
   c. urcaryotes.
   d. nocaryotes.

8. The concept that human and animal diseases are caused by microorganisms is called the
   a. cell theory.
   b. germ theory.
   c. causative theory.
   d. disease theory.

9. Cells with a relatively complex morphology that have a true
   membrane-delimited nucleus are called
   a. procaryotes.
   b. eucaryotes.
   c. urcaryotes.
   d. nocaryotes.

10. Who, of the following, showed that mutations in bacteria were random and not directed by the
    environment?
    a. Beadle & Tatum
    b. Avery, MacLeod, & McCarty
    c. Luria & Delbruck
    d. Hershey & Chase

11. T F Microorganisms are useful in bioremediation processes to reduce the effects of pollution.

12. T F   Genetically engineered organisms have not yet proven useful in providing products for
    human consumption.

13. The work delineating the relationship between genes and enzymes was done using which of the
    following types of microorganism?
    a. bacteria
    b. fungi
    c. viruses
    d. algae

14. T F Microorganisms probably constitute the largest single component of the earth's biomass.

15. Antiseptic surgery was pioneered by
    a. Pasteur.
    b. Lister.
    c. Jenner.
    d. Kitasato.

16. Emil von Behring and Shibasaburo Kitasato demonstrated the body's defense
    mechanism against disease known as
    a. humoral immunity.
    b. cell-mediated immunity.
    c. both humoral immunity and cell-mediated immunity.
    d. neither humoral immunity nor cell-mediated immunity.

17. The first surgical antiseptic to be used was
    a. iodine.
    b. ethanol.
    c. phenol.
    d. None of these.

18. Older cultures of bacteria that have lost their ability to cause disease
    are said to be
    a. impotent.
    b. virulent.
    c. pathogenic.
    d. attenuated.

19. Who of the following is credited with developing and documenting the
    first vaccination procedure against smallpox?
    a. Koch
    b. Pasteur
    c. Jenner
    d. Lister

20. Who of the following is credited with developing a vaccine against
    chicken cholera?
    a. Koch
    b. Pasteur
    c. Jenner
    d. Lister

21. Who of the following first proposed that yeast cells could convert sugar
    to alcohol?
    a. Pasteur
    b. Tyndall
    c. Schwann
    d. Busch

22. Elie Metchnikoff discovered ___________, which is the basis for
    cell-mediated immunity.

23. An Italian physician, ___________, challenged the concept of spontaneous generation by
    demonstrating that maggots did not arise from decaying meat but rather from developing fly eggs.

24. ___________ discovered that soil bacteria could oxidize iron, sulfur, and ammonia to obtain energy.

25. ___________ isolated a root nodule bacterium capable of nitrogen fixation.

26. The branch of microbiology that deals with diseases of humans and animals is called ___________
    microbiology.

27. The branch of microbiology that deals with the mechanisms by which the human body protects itself
    from disease-causing organisms is called ________________.

28. ___________ microbiology endeavors to control the spread of communicable diseases.

29. The branch of microbiology that studies the relationship between microorganisms and their habitats is
    called ___________.

30. __________ and __________ microbiology investigates the spoilage of products for human consumption and the use of microorganisms in the production of cheese, yogurt, pickles, beer, and the like.

31. __________ microbiology involves the use of microorganisms to make products such as antibiotics, vaccines, steroids, alcohols, vitamins, amino acids, and enzymes.

32. Microbial __________ investigates the synthesis of antibiotics and toxins, the production of energy with microorganisms, and the ways in which microorganisms survive harsh environmental conditions.

33. Microbial __________ focuses on the nature of heredity and how it regulates the development and function of cells and organisms.

34. The insertion of new genes into organisms in order to investigate their functions or to produce useful organisms with new functions is called __________.

35. T F That yeast cells were capable of converting sugar to alcohol was first proposed by Theodor Schwann.

36. T F M. J. Berkeley demonstrated that the great potato blight of Ireland was caused by a fungus.

37. T F Invisible living creatures were thought to exist and cause disease long before they were ever observed.

38. T F Mutations are directed by the environment.

39. T F Because they are not classified as living organisms, viruses are not generally studied by microbiologists.

40. Who of the following defined the relationship between genes and enzymes?
    a. Luria and Delbruck
    b. Avery, MacLeod, and McCarty
    c. Beadle and Tatum
    d. Hershey and Chase

41. T F The criteria for establishing a causative link between a particular microorganism and a particular disease were first *developed* by Robert Koch.

42. T F The criteria for establishing a causative link between a particular microorganism and a particular disease were first *published* by Robert Koch.

43. T F Edward Jenner was the first to use what we now call vaccination in order to prevent disease.

44. T F Gelatin is better than agar for use in solidifying biological growth media because it also serves as a fermentable substrate for the growing organisms.

45. T F The criteria for establishing a causative link between a particular microorganism and a particular disease were first developed by Jacob Henle.

46. T F The criteria for establishing a causative link between a particular microorganism and a particular disease were first published by Jacob Henle.

47. T F Agar is used as a solidifying agent for microbiological media because it is not readily digested by microorganisms.

48. T F Charles Chamberland discovered viruses and their role in disease by using porcelain filters that he constructed to retain bacteria.

49. T F The first disease to be identified as being caused by a virus was anthrax.

50. T F Although Edward Jenner was not the first to use vaccination procedures to prevent disease, he was the first to document such procedures.

51. T F John Tyndall demonstrated that dust particles in the air carried microorganisms and hat sterile broth remained sterile if dust was kept out of flasks containing the broth.

52. T F Agastino Bassi demonstrated that silkworm disease was caused by a fungus.

53. T F Robert Koch developed a vaccine that could be used to prevent anthrax.

54. T F The relationship between genes and enzymes was established by Luria and Delbruck.

55. T F Scientists trained in microbiology have been instrumental in the elucidation of the genetic code.

56. T F Microbiologists study a variety of organisms, but all are considered members of the kingdom Monera.

57. Microbiologists who study algae are referred to as __________.

58. Microbiologists who study fungi are referred to as __________.

59. The use of enrichment cultures and selective media was pioneered by
    a. Beijerinck
    b. Winogradsky
    c. Both a and b are correct.
    d. Neither a nor b is correct.

60. Alternatives to the 5 kingdom classifications have been devised that divide living organisms into 2 or 3 __________ with multiple kingdoms contained within.

Chapter 2 Sample Test Questions on The Study of Microbial Structure: Microscopy and Specimen Preparation

1. Acid-fast organisms such as *Mycobacteriumt tuberculosis* resist decolorization by an acid-alcohol wash because of the high concentration of __________ in their cell walls.
    a. proteins
    b. carbohydrates
    c. lipids
    d. peptidoglycan

2. When smears of microorganisms are stained, the smear is heat-fixed in order to
   a. kill the organism.
   b. attach the organism firmly to the slide.
   c. kill the organism and attach the organism firmly to the slide.
   d. neither kill the organism nor attach the organism firmly to the slide.

3. Small internal cell structures are best visualized with a
   a. light microscope.
   b. dark-field microscope.
   c. transmission electron microscope.
   d. scanning electron microscope.

4. In transmission electron microscopy, spreading a specimen out in a thin film with uranyl acetate, which does not penetrate the specimen, is called
   a. freeze-etching.
   b. simple staining.
   c. shadow staining.
   d. negative staining.

5. Scanning electron microscopy is most often used to reveal
   a. surface structures.
   b. internal structures.
   c. both surface and internal structures simultaneously.
   d. either surface or internal structures, but not simultaneously.

6. As the magnification of a series of objective lenses increases, the working distance
   a. increases.
   b. decreases.
   c. stays the same.
   d. cannot be predicted.

7. The composition of the ___________ is the basis of the differential reaction of bacteria to the Gram-staining procedure.
   a. chromosome
   b. cytoplasmic membrane
   c. cell wall
   d. slime layer

8. The instrument that produces a bright image of the specimen against a dark background is called a(n) ___________ microscope.
   a. phase-contrast
   b. electron
   c. bright-field
   d. dark-field

9. An instrument that converts slight differences in refractive index and density of cell structures is called a(n) ___________ microscope.
   a. phase-contrast
   b. electron
   c. fluorescence
   d. densitometric

10. Immersion oil is used with some microscope lenses because it has the same __________ as glass.
    a. optical density
    b. refractive index
    c. optical density and refractive index
    d. neither optical density nor refractive index

11. A microscope that has the property by which the objective lenses can be changed without losing focus on the specimen is said to be
    a. equifocal.
    b. totifocal.
    c. parfocal.
    d. optifocal.

12. A microscope that exposes specimens to ultraviolet, violet, or blue light and forms an image with the resulting light emitted at a different wavelength is called a __________ microscope.
    a. phase-contrast
    b. dark-field
    c. scanning electron
    d. fluorescence

13. A 45× objective and a 10× ocular produce a total magnification of
    a. 35×.
    b. 55×.
    c. 450×.
    d. 4.5×.

14. A 30× objective and a 20× ocular produce a total magnification of
    a. 230×.
    b. 320×.
    c. 50×.
    d. 600×.

15. A 40X objective and a 15× ocular produce a total magnification of
    a. 30×.
    b. 600×.
    c. 3×.
    d. 60×.

16. A 100× objective and a 15× ocular produce a total magnification of
    a. 115×.
    b. 150×.
    c. 1,500×.
    d. 1,000×.

17. In the Gram-staining procedure, the primary stain is
    a. iodine.
    b. safranin.
    c. crystal violet.
    d. alcohol.

18. In the Gram-staining procedure, the decolorizer is
    a. iodine.
    b. safranin.
    c. crystal violet.
    d. alcohol.

19. In the Gram-staining procedure, the secondary stain is
    a. iodine.
    b. safranin.
    c. crystal violet.
    d. alcohol.

20. In the Gram-staining procedure, the mordant is
    a. iodine.
    b. safranin.
    c. crystal violet.
    d. alcohol.

21. After the primary stain has been added but before the decolorizer has been used, gram-positive
    organisms are stained ___________ and gram-negative organisms are stained __________.
    a. blue; blue
    b. blue; colorless
    c. blue; pink
    d. pink; pink

22. After the decolorizer has been added, gram-positive organisms are stained __________ and gram-
    negative organisms are stained __________.
    a. blue; blue
    b. blue; colorless
    c. blue; pink
    d. pink; pink

23. After the secondary stain has been added, gram-positive organisms are   stained __________ and
    gram-negative organisms are stained __________.
    a. blue; blue
    b. blue; colorless
    c. blue; pink
    d. pink; pink

24. If the decolorizer is left on too long in the Gram-staining procedure, gram-positive organisms will be
    stained __________ and gram-negative organisms will be stained __________.
    a. blue; blue
    b. blue; colorless
    c. blue; pink
    d. pink; pink

25. If the decolorizer is not left on long enough in the Gram-staining procedure, gram-positive organisms will be stained ___________ and gram-negative organisms will be stained __________.
    a. blue; blue
    b. blue; colorless
    c. blue; pink
    d. pink; pink

26. Which of the following is considered to be a differential staining procedure?
    a. Gram stain
    b. Acid-fast stain
    c. Both Gram stain and Acid-fast stain.
    d. Neither Gram stain nor Acid-fast stain.

27. T F Immersion oil is used with the 100× objective lens in order to increase the amount of light passing through a specimen.

28. T F The Gram-staining procedure is one of the most widely used differential stains because it allows rapid identification of a microorganism with little additional testing.

29. T F Because transmission electron microscopy uses electrons rather than light, it is not necessary to stain biological specimens before observing them.

30. T F It is possible to build a light microscope capable of 10,000× magnification, but the image would not be sharp because resolution is independent of magnification.

31. T F Immersion oil is used to prevent a specimen from drying out.

32. T F Light is refracted (bent) whenever it crosses the interface between two materials with different refractive indices.

33. T F Resolution decreases when the wavelength of the illuminating light decreases.

34. T F The coarse focusing adjustment knob should be used only with the high power objective.

35. T F Gram staining divides bacterial species into roughly two equal groups.

36. T F Negative staining is widely used to visualize capsules which are intensely stained by the procedure.

37. The ___________ is the distance between the center of a lens and the point at which it focuses parallel beams of light.

38. The ___________ is the distance between the specimen and the objective lens when the specimen is in focus.

39. The ___________ is the point at which a lens focuses parallel beams of light.

40. An enlarged image formed within the microscope by the objective lens is called the ___________ image.

41. The __________ image is formed by the ocular lens and appears to lie just beyond the stage about 25 cm away.

42. An electron microscope uses __________ to focus beams of electrons onto a specimen.

43. The special dyes used in fluorescence microscopy that absorb light at one wavelength and emit light at a different wavelength are called __________.

44. T F Negative staining is frequently used to reveal the presence of diffuse capsules surrounding some bacteria.

45. Denser regions of the specimen scatter __________ electrons and, therefore, appear __________ in the image projected onto the screen of a transmission electron microscope.
    a. more; lighter
    b. more; darker
    c. fewer; darker
    d. fewer; lighter

46. The procedure in which a single stain is used to visualize microorganisms is called __________ staining.

47. __________ involves the use of heavy metals cast at a 45-degree angle from horizontal to create a three-dimensional image for electron microscopy.

48. __________ is the process by which internal and external structures of cells and organisms are preserved and maintained in position.

49. Thin films of bacteria that have been air-dried onto a glass microscope slide are called __________.

50. In order to view a specimen with a total magnification of 100× a __________ objective must be used if the ocular is 10×.

51. A procedure that divides organisms into two or more groups depending on their individual reactions to the same staining procedure is referred to as __________ staining.

52. __________ breaks frozen specimens along lines of greatest weakness, usually down the middle of membranes so that they may be observed by scanning electron microscopy.

53. T F A mordant is a substance that increases the reaction between a cell and a dye so that the cell is more intensely stained.

54. T F In order to stain flagella so that they may be readily observed, it is usually necessary to increase their thickness.

55. T F It was possible to view chromosomes only after the invention of the electron microscope because they are too small to be seen with a light microscope.

56. T F The substage condenser is used to focus light onto the specimen in order to increase the visualization of the image.

57. T F The scanning tunneling electron microscope creates a three-dimensional image of the specimen with a resolution such that individual atoms can be observed.

58. An atomic force microscope uses a scanning probe that maintains a fixed distance from the surface of the specimen. It is useful for specimens that
    a. do not conduct electricity well
    b. have extremely uneven surfaces
    c. Both a and b are correct.
    d. Neither a nor b is correct.

Answers to Sample Test Questions from Video 1

1. Biomass
2. ecosystems
3. C
4. E
5. E
6. False
7. B
8. B
9. C
10. A
11. True
12. True
13. False
14. D
15. genome

Answers to Sample Test Questions from Text Chapter 2 (The Study of Microbial Structure: Microscopy and Specimen Preparation) Web Site

1. True
2. True
3. Focuses light on the specimen
4. 400
5. Phase contrast microscopy
6. Surface morphology
7. True
8. Transmission electron microscope
9. 0.9 microns
10. Does not destroy internal structures

Answers to Chapter 1 Sample Test Questions on The History and Scope of Microbiology

The question in this test bank have been rated for difficulty. The ratings run from 1 (for the easiest) to 8 (for the most difficult).

#	Answer / Diff.
1	C / 2
2	B / 2
4	A / 2
5	B / 5
6	C / 5
7	A / 3
8	B / 4
9	B / 3
10	C / 4
11	T / 3
12	F / 3
13	B / 5
14	T / 4
15	B / 5
16	A / 6
17	C / 7
18	D / 7
19	C / 5
20	B / 5
21	C / 8
22	phagocytosis / 5
23	Francesco Redi
24	Sergius Winogradsky / 7
25	Martinus Beijerinck / 7
26	medical / 5
27	immunology / 5
28	Public health / 5
29	microbial ecology / 5
30	Food; dairy / 5
31	Industrial / 5
32	physiology / 5
33	genetics
34	genetic engineering
35	T/7
36	T / 5

#	Answer / Diff.
37	T / 5
38	F / 6
39	F / 5
40	C / 5
41	F / 8
42	T / 5
43	F / 8
44	F / 7
45	T/8
46	F/8
47	T/5
48	F/9
49	F/5
50	T/8
51	T/5
52	T/5
53	F/6
54	F/5
55	T/5
56	F/4
57	phycologists/7
58	mycologists/7
59	C/6
60	empires (domains)/6

Answers to Chapter 2 Sample Test Questions on The Study of Microbial Structure: Microscopy and Specimen Preparation

The question in this test bank have been rated for difficulty. The ratings run from 1 (for the easiest) to 8 (for the most difficult). Questions rated 7 or 8 may be most effective when used as challenge questions.

#	Answer/ Diff.		#	Answer/ Diff.
1	C / 6		43	fluorochromes/5
2	C / 3		44	T/5
3	C / 5		45	A/7
4	D / 6		46	simple/2
5	A / 5		47	shadowing/6
6	B / 7		48	fixation/4
7	C / 5		49	smears/3
8	D / 5		50	10X/4
9	A / 5		51	differential/5
10	B / 8		52	freeze-fracturing/5
11	C / 5		53	T/3
12	D / 5		54	T/5
13	C / 5		55	F/4
14	D / 5		56	T/3
15	B / 5		57	T/8
16	C / 5		58	A/7
17	C / 4			
18	D / 4			
19	B / 4			
20	A / 4			
21	A / 5			
22	B / 5			
23	C / 5			
24	D / 7			
25	A / 7			
26	C / 5			
27	T / 8			
28	F / 5			
29	F / 8			
30	T / 9			
31	F / 6			
32	T / 6			
33	F / 8			
34	F / 3			
35	T / 2			
36	F/6			
37	focal length/5			
38	working distance/7			
39	focal point/4			
40	real/8			
41	virtual/8			
42	magnets/4			

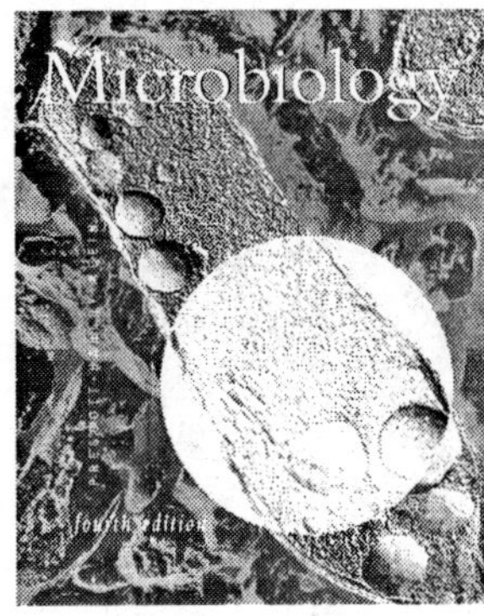

Microbial Cell Biology
Program 2
The Unity of Living Systems

> In the field of observation,
> chance favors only prepared minds.
> – Louis Pasteur

One of microbiology's fifty most significant events in the past 125 years:

1885

Louis Pasteur injected a child, Joseph Meister, with "aged" spinal cord infected with rabies virus. Pasteur used the term "virus," meaning poison, but had no idea of the nature of the causative organism. Although the treatment was successful, the experiment itself was an ethical violation of research standards. Pasteur new he was giving the child successively more dangerous portions.

Student Assignments

- **Watch** Video 2: The Unity of Living Systems
- **Read** Chapter 3: Procaryotic Cell Structure and Function
    Chapter 4: Eucaryotic Cell Structure and Function , and
    Chapter 16: The Viruses: Introduction and general Characteristics
    in *Microbiology*, 4th ed, Prescott/Harley/Klein, McGraw-Hill
- **Lab Experiments** from *Laboratory Exercises in Microbiology*, 4th ed., Harley/Prescott
    McGraw-Hill
    Exercise 6: The Hanging Drop Slide and Bacterial Motility
    Exercise 7: Negative Staining
    Exercise 8: Smear Preparation and Simple Staining
    Exercise 9: Gram Stain
    Exercise 10: Acid-Fast Staining (Ziel-Neelsen and Kinyoun) Procedures
    Exercise 11: Endospore Staining
    Exercise 12: Capsule Staining
    Exercise 13: Flagella Staining
    Exercise 57: Isolation of *Escherichia coli* bacteriophages from Sewage and
        Determining Bacteriophage Titers

Lesson Overview

Video 2: The Unity of Living Systems
All cellular organisms–procaryotic and eukaryotic–share basic chemical similarities. Out of these similarities, however, emerge diverse patterns of cell assembly. Students encounter the tools they need to understand various cell types and their relationships to noncell entities such as viruses.

Chapter 3: Procaryotic Cell Structure and Function
This chapter provides a description of the procaryotic cell, beginning with the general features of size, shape, and arrangement. Then the general features of biological membranes and the specific features of procaryotic membranes are given. Important internal structures of procaryotes, such as the cytoplasmic matrix, the ribosomes, the inclusion bodies, and the nucleoid are described, in addition to structures external to the cell, such as the cell wall, capsule, pili, and flagella. The differences between the cell walls of gram-positive organisms and gram-negative organisms are discussed and the mechanism of this differential staining reaction is explained. The chapter concludes with a discussion of bacterial chemotaxis and bacterial endospores.

Chapter 4: Eucaryotic Cell Structure and Function
This chapter focuses on eucaryotic cell structure and function. Although procaryotic organisms are immensely important in microbiology, eucaryotic microorganisms such as fungi, algae, and protozoa are also prominent members of many ecosystems, and some have medical significance as etiological agents of disease as well. The chapter concludes with a comparison of eucaryotic and procaryotic cells.

Chapter 16: The Viruses: Introduction and General Characteristics
Viruses are generally small, acellular entities that possess only a single type of nucleic acid and that must use the metabolic machinery of a living host in order to reproduce. Viruses have been and continue to be of tremendous importance for a variety of reasons: many human diseases have a viral etiology; the study of viruses has contributed greatly to our knowledge of molecular biology; and the blossoming field of genetic engineering is largely based on discoveries in the field of virology. This chapter focuses on the general properties of viruses, the development of the science of virology, and the methodology used in the study of virology.

Learning Objectives

After successfully completing this unit, the student will be able to:

- describe the various sizes, shapes, and cellular arrangements exhibited by bacteria
- describe the bacterial plasma membrane and the limited internal membrane structures found in procaryotes
- describe the appearance, composition, and function of the various internal structures found in procaryotic organisms (such as inclusion bodies, ribosomes, and the nucleoid)
- define the composition of gram-positive and gram-negative cell walls and explain how these differences contribute to the differential reaction to the Gram-staining procedure

- describe external structures such as capsules, fimbriae and flagella
- diagram and describe the various arrangements of bacterial flagella
- describe how bacteria use their locomotive ability to swim toward chemical attractants and away from chemical repellents
- describe the production of the bacterial endospore and how it enables spore-forming bacteria to survive harsh environmental conditions and renew growth when the environment becomes conducive to growth
- discuss the various elements of the cytoskeleton (microfilaments, intermediate filaments, microtubules) with regard to their structure and various functions within the cell
- discuss the composition, structure, and function of each of the internal organelles, such as the endoplasmic reticulum, Golgi apparatus, lysosomes, ribosomes, mitochondria, chloroplasts, nucleus, and nucleolus
- discuss the mechanism of endocytosis and the difference between phagocytosis and pinocytosis
- compare mitosis and meiosis
- compare and contrast procaryotes and eucaryotes
- define viruses and discuss the implications of the concepts embodied in the definition
- discuss the various requirements for culturing viruses
- discuss the methodology employed for virus purification and enumeration
- discuss the composition and arrangement(s) of viral capsids
- discuss the variety found in viral genomes (DNA or RNA, single or double stranded, linear or circular, etc.)
- describe the way in which viruses are classified

Course Integrator

Chapter 3
An Overview of Procaryotic Cell Structure (p.37)

Hyperclinic	Diagnostics: Direct Examination, Direct Detection
Hyperclinic	Diagnostics: Culture
	Identibacter interactus
Web Link	*Microbiology* Online Learning Center: http://www.mhhe.com/biosci/cellmicro/prescott/
Web Link	CELLS Alive! (http://www.cellsalive.com)
Video *Unseen Life on Earth*	Unit #2 The Unity of Living Things

Procaryotic Cell Membranes (p.40)

Microbes in Motion II	Bacterial Structure & Function/Cell Membrane
Transparency	Gram-positive Cell Morphology Figure 3.4
Transparency	Plasma Membrane Structure Figure 3.7
Visual Resource Library	Plasma Membrane Structure Figure 3.7

The Cytoplasmic Matrix (p.44)

Visual Resource Library	Cross-section of *E. coli* Figure 3.10
Visual Resource Library	Chaperones and protein folding Figure 3.16
Transparency	Chaperones and Protein Folding Figure 3.16

The Nucleoid (p.50)

The Procaryotic Cell Wall (p.51)

Visual Resource Library	Bacterial Cell Wall Figure 3.19
Visual Resource Library	Peptidoglycan Subunit Figure 3.20
Transparency	Peptidoglycan Subunit Figure 3.20
Visual Resource Library	Peptidoglycan Structure Figure 3.23
Transparency	Peptidoglycan Structure Figure 3.23
Visual Resource Library	peptidoglycan Cross-Links Figure Number 3.22
Visual Resource Library	Gram-positive Envelope Figure 3.25
Transparency	Gram-positive Envelope Figure 3.25
Visual Resource Library	Gram-negative Envelope Figure 3.27
Transparency	Gram-negative Envelope Figure 3.27
Visual Resource Library	Lipopolysaccharide Structure Figure 3.29
Transparency	Lipopolysaccharide Structure Figure 3.29
Microbes in Motion II	Bacterial Structure & Function
/Gram Positive Cell	
/Gram Negative Cell	
/Cell Wall	
Microbes in Motion II	Antimicrobial Action/Cell Wall Inhibitors

Chapter 4
An Overview of Eucaryotic Cell Structure (p. 74)

Visual Resource Library	Eucaryotic Cell Ultrastructure Figure 4.3
Transparency	Eucaryotic Cell Ultrastructure Figure 4.3
Microbes in Motion II	Mycology/General Eucaryotic Structures
Hyperclinic	Diagnostics: Direct Examination, Direct Detection, *Hyperclinic*
Diagnostics: Culture	
Web Link	*Microbiology* Online Learning Center:
	http://www.mhhe.com/biosci/cellmicro/prescott/
Web Link	Virtual Cell: http://ampere.scale.uiuc.edu/~m-lexa/cell/cell.html
Web Link	CELLS Alive!: http://www.cellsalive.com
Video *Unseen Life on Earth*	Unit #2 The Unity of Living Things

The Cytoplasmic Matrix, Microfilaments, Intermediate Filaments, and Microtubules (p.74)
The Endoplasmic Reticulum (p.77)
The Golgi Apparatus (p.77)

Lysosomes and Endocytosis (p.78)

Visual Resource Library	Lysosome Formation and Function Figure 4.9a
Visual Resource Library	The Vacuome Figure 4.10

Eucaryotic Ribosomes (p.80)

Mitochondria (p.81)

Visual Resource Library   The Mitochondrion Figure 4.13

Chloroplasts (p.81)
Visual Resource Library   The Chloroplast Figure 4.15

The Nucleus and Cell Division (p.83)
Visual Resource Library   The Cell Cycle Figure 4.18
Visual Resource Library   Mitosis and Cytokinesis Figure 4.19
Web Link   The stages of mitosis. A tour with pictures:
http://www.inform.umd.edu/PBIO/wolniak/wolniakmitosis.html

External Cell Coverings (p.87)

Cilia and Flagella (p.89)
Visual Resource Library   Cilia and Flagella Structure Figure 4.27

Comparison of Procaryotic and Eucaryotic Cells (p.90)

Chapter 16
Early Development of Virology (p.336)
Web Link   *Microbiology* Online Learning Center
http://www.mhhe.com/biosci/cellmicro/prescott/

General Properties of Viruses (p. 337)
Web Link   All the Virology on the Web
(http://www.tulane.edu/~dmsander/garryfavweb.html)
Microbes in Motion II   Virology/Viral Structure/Virus Taxonomy

The Cultivation of Viruses (p.337)
Microbes in Motion II   Virology/Viral Detection/Culture
Microbes in Motion II   Virology/Viral Pathogenesis/Host Cell Damage

Virus Purification and Assays (p. 339)
Visual Resource Library   Gradient Centrifugation Figure 16.7
Hyperclinic   Diagnostics/Nucleic Acid Detection
Hyperclinic   Diagnostics/Antigen Detection
Hyperclinic   Antibody Detection

The Structure of Viruses (p. 342)
Microbes in Motion II   Virology/Viral Structure/Virus Structure and Shape
Transparency   Tobacco Mosaic Virus Structure Figure 16.11 b, c
Visual Resource Library   Tobacco Mosaic Virus Structure Figure 16.11 b, c
Transparency   Icosahedral Capsid Structure Figure 16.13
Visual Resource Library   Icosahedral Capsid Structure Figure 16.13
Transparency   Influenza Virus Figure 16.17b
Visual Resource Library   Influenza Virus Figure 16.17b

Transparency	T4 Coliphage Structure Figure 16.19a
Visual Resource Library	T4 Coliphage Structure Figure 16.19a
Visual Resource Library	Vaccinia Virus Morphology Figure 16.18a
Web Link	The Big Picture Book of Viruses (http://www.tulane.edu/~dmsander/Big-Virology/BVHomePage.html)
Web Link	Electron Microscopic Images of Human Viruses (http://www.vct.ac.za/depts/mmi/stannard/linda.html)

Principles of Virus Taxonomy (p. 352)

Microbes in Motion II	Virology/Viral Structure/Virus Taxonomy
Web Link	The Big Picture Book of Viruses (http://www.tulane.edu/~dmsander/Big-Virology/BVHomePage.html)
Web Link	The Index Virum (International Committee on Taxonomy

Microbial Terms Used in Video 2 in Order of Appearance with Page Indexing to the Textbook (*Microbiology*, 4th ed., Prescott/Harley/Klein, McGraw-Hill)

Salmonella 477	chloroplasts 81	common cold 740
Staphylococcus 491	nucleus 83	chickenpox 739
E. coli 477	cell wall 208	measles 741
procaryotic cells 90	slime layer 58	viruses 335
eukaryotic cells 83	cell membrane 87	envelope 51
lysosomes 78	flagella 89	capsid 343
mitochondria 81	endosymbiosis 83, 395	energy 163
ribosomes 80	symbiosis 566, 892	metabolism 164

Web Links Related to Video 2 (The Unity of Living Systems) and Readings in Chapter 3 (Procaryotic Cell Structure and Function) (Go to www.mhhe.com/prescott, Student on Line Learning Center, Double Click on Chapter 3, Double Click on Web Site)

Cells Alive
http://www.cellsalive.com

Gram Stain. A teaching page on the gram stain where you can view gram-stained images of bacteria that are medically important.
http://www.meddean.luc.edu/lumen/DeptWebs/microbio/med/gram/gram-stn.htm

Biological Identity of Procaryotes. Kenneth Todar, University of Wisconsin–Madison, Department of Bacteriology.
http://www.bact.wisc.edu/Bact303/TheProcaryotes

The Diversity of Metabolism in Procaryotes. Kenneth Todar, University of Wisconsin–Madison, Department of Bacteriology.
http://www.bact.wisc.edu/Bact303/bact303metabolism

Web Links Related to Video 2 (The Unity of Living Systems) and Readings in Chapter 4 (Eucaryotic Cell Structure and Function)

The Stages of Mitosis. Stephen Wolniak, Department of Cell Biology and Molecular Genetics, University of Maryland.
http://www.inform.umd.edu/PBIO/wolniak/wolniakmitosis.html

Virtual Cell. This site is a collection of still images, texts, and movies covering the structure and functioning of a typical plant cell.
http://ampere.scale.uiuc.edu/~m-lexa/cell/cell.html

The Cell Nucleus. Cell biology graduate program at the University of Texas medical branch.
http://cellbio.utmb.edu/cellbio/nucleus.htm

Yahoo! Science. Cell Biology. More links to cell biology sites for you to explore.
http://dir.yahoo.com/Science/Biology/Cell_Biology/

Web Links Related to Video 2 (The Unity of Living Systems) and Readings in Chapter 16 (The Viruses: Introduction and General Characteristics)

All the Virology on the WWW. This site seeks to be the best single site for virology information on the Internet. All of the virology related web sites have been collected.
http://www.tulane.edu/~dmsander/garryfavweb.html

The Big Picture Book of Viruses. This site is intended to serve as both a catalog of virus pictures on the WWW and as an educational resource to those seeking more information about viruses.
http://www.tulane.edu/~dmsander/Big_Virology/BVFamilyGenome.html

Electron Microscopic Images of Human Viruses. Linda Stannard's illustrated tutorial on the morphology of most of the clinically significant viruses. The section on the Hepatitis B virus is especially recommended.
http://www.uct.ac.za/depts/mmi/stannard/linda.html

Visualization of Viruses at the University of Wisconsin–Madison.
http://www.bocklabs.wisc.edu/virusviztop.html

The Index Virum. This site presents a list of virus taxa that reflect the currently approved classification of the International Committee on taxonomy of Viruses (ICTV).
http://life.anu.edu.au/viruses/Ictv/index.html

The International Committee on Taxonomy of Viruses. From the National center for Biotechnology.
http://www.ncbi.nlm.nih.gov/ICTV/

Microbial Terms Used in Chapter 3 (Procaryotic Cell Structure and Function)

Microbial Terms Used in Chapter 4 (Eucaryotic Cell Structure and Function)

autophagic vacuole 78
axoneme 90
basal body 90
cell cycle 85
cell wall 87
chloroplast 81
chromatin 83
chromosome 83
cilia 89
cisternae 77
cristae 81
cytokinesis 87
cytoplasmic matrix 74
cytoskeleton 76
dictyosome 77
diploid 87
dynein 90
endocytosis 78
endoplasmic reticulum
(ER) 77

endosome 78
endosymbiotic theory
83
eucaryotic cells 90
F1 particle 81
flagella 89
Golgi apparatus 77
grana 83
haploid 87
intermediate filament
76
interphase 85
lysosome 78
meiosis 87
microfilament 75
microtubule 76
mitochondrion 81
mitosis 85
nuclear envelope 84
nuclear pores 84
nucleolus 84

nucleus 83
organelle 74
pellicle 89
phagocytosis 78
pinocytosis 78
plastid 81
polyribosomes 80
polysomes 80
primary lysosomes 78
procaryotic cells 93
pyrenoid 83
residual body 78
ribosomal RNA 85
rough or granular ER
77
secondary lysosomes
78
smooth or agranular
ER 77
stroma 83
thylakoid 83

Microbial Terms Used in Chapter 16 (The Viruses: Introduction and General Characteristics)

bacteriophage 338
binal symmetry 349
capsid 343
capsomeres 344
complex virus 343
cytopathic effects 338
differential
centrifugation 339
envelope 343
gradient centrifugation
339
helical 343
hemagglutination assay
342

hexamers (hexons) 344
isoahedral 343
infectious dose 342
lethal dose 342
minus or negative
strand 346
necrotic lesion 338
nucleocapsid 343
pentamers (pentons)
344
phage 38
plaque 338
plaque assay 342

plaque-forming units
342
plus strand or positive
strand 346
protomers 344
segmented genome
346
spike or peplomer 349
viron 337
virologist 336
virology 336
virus 337

Sample Test Questions from Video 2 (The Unity of Living Systems)

1. Do humans have anything in common with microbes such as *Salmonella, E. coli,* and *Staphylococcus*?
   a. Yes
   b. No

2. The proteins and nucleic acids found in bacteria, fungi, plants, and animals are all different.
   a. True
   b. False

3. Which of the following is not an eucaryotic cell?
   a. Plants
   b. Animals
   c. Bacteria
   d. Viruses
   e. Both c and d.

4. Plants, fungi, and animals share the procaryotic cell plan.
   a. True
   b. False

5. Procaryotic cells have more component parts than do eucaryotic cells.
   a. True
   b. False

6. Most procaryotic cells have a cell wall.
   a. True
   b. False

7. Which of the following would you not find in a procaryotic cell?
   a. Cell wall
   b. Slime layer
   c. Membrane enclosed nucleus
   d. Ribosomes

8. The structure that some bacteria used for motility is a __________.

9. The process whereby mitochondria and chloroplasts took up residence inside primitive eucaryotic cells is called __________.

10. Our intestines contain very few microbes.
    a. True
    b. False

11. Which of the following human diseases is not caused by a virus?
    a. The common cold.
    b. Measles
    c. Chicken pox
    d. Influenza
    e. Strep throat

12. Viruses are one type of microbial cell.
    a. True
    b. False

13. The outermost layer of a virus is called an __________.

14. Viruses can replicate on their own.
    a. True
    b. False

15. Viruses multiply by taking over the cells metabolic machinery.
    a. True
    b. False

16. In all cells, the blueprint for genetic information is in the form of __________.

17. Based on the work of Dr. Craig Venter, every gene in the human body can be traced to other species–such as microbes.
    a. True
    b. False

18. Colon cancer is a genetic disorder that can be traced to the *E. coli* bacteria that live in our intestines.
    a. True
    b. False

19. Each genome is a recorded history of life.
    a. True
    b. False

20. The term that describes the chemical reactions that occur inside an organism is __________.

Sample Test Questions from Text Chapter 3 (Procaryotic Cell Structure and Function) Web Site

1. The significance of the plasma membrane is that
    O it selectively allows some molecules to pass into the organism
    O it prevents movement of molecules out of the organism
    O it is the site of protein synthesis
    O all of the above

2. The plasma membrane and most other membranes are composed of a lipid bilayer in which integral proteins are buried.
    O True
    O False

3. Procaryotic cells are usually smaller than eucaryotic cells.
   O True
   O False

4. The most commonly encountered bacteria are roughly spherical. The microbiological term describing this shape is
   O coccus
   O bacillus
   O pleomorphic

5. Another common bacterial shape is that of a rod, often called
   O coccus
   O bacillus
   O pleomorphic

6. In bacterial cells, ribosomes are packed into the cytoplasmic matrix and also loosely attached to the plasma membrane. What is the function of ribosomes?
   O Site of energy production
   O Site of protein synthesis
   O Site of genetic reproduction

7. What is a plasmid?
   O Circular, double-stranded DNA particles
   O Circular, single-stranded RNA particles
   O A bacterial chromosome

8. Plasmids are important to the genetics of many bacteria. This is because
   O they are inherited from one generation to the next
   O they may carry genes that give their host a selective advantage
   O they can render bacteria drug-resistant
   O all of the above

9. Bacteria do not always swim aimlessly but are attracted by such nutrients such as sugar and amino acids, and are repelled by harmful substances and bacterial waste products. Movement toward chemical attractants and away from repellents is called
   O gliding motility
   O tumbling
   O chemotaxis

10. Most procaryotes have strong cell walls that give them shape.
   O True
   O False

Sample Test Questions from Text Chapter 4 (Eucaryotic Cell Structure and Function) Web Site

1. Eucaryotic cells differ most obviously from procaryotic cells in having complex membranous organelles in the cytoplasmic matrix and most of their genetic material in a membrane-delimited nucleus.
   O True
   O False

2. In eucaryotes, genetic material is distributed between cells by the highly organized, complex processes called mitosis and meiosis.
   - O True
   - O False

3. Despite great morphological difference between eucaryotic cells and prokaryotes, they are remarkably similar metabolically.
   - O True
   - O False

4. The endosymbiotic theory states that eucaryotic organelles such as mitochondria and chloroplasts arose when bacteria established a symbiotic relationship with the eucaryotic ancestor and then evolved into organelles.
   - O True
   - O False

5. In contrast to procaryotes, many eucaryotic microorganisms do not have a cell wall. When they do, the walls of eucaryotic cells are generally constructed of polysaccharides and are simpler than that of procaryotic cells.
   - O True
   - O False

6. The diploid state of a microorganism refers to the times when cells contain _______ copy (ies) of each chromosome and occurs after mitotic cell division.
   - O one
   - O two
   - O three

7. Cilia and flagella are the most prominent structures that enable eucaryotic cells to move.
   - O True
   - O False

8. Cell walls, rather cytoskeletons, give eucaryotic cells their shape.
   - O True
   - O False

9. Some examples of eucaryotic organisms are algae, fungi, protozoa and actinomycetes.
   - O True
   - O False

10. The mitochondria is the site of
    - O protein synthesis
    - O lipid synthesis
    - O energy production

Sample Test Questions from Text Chapter 16 (The Viruses: Introduction and General Characteristics) Web Site

1. Viruses are simple, acellular, infectious agents.
   - O True
   - O False

2. Viruses can reproduce independently from the host cell.
   - O True
   - O False

3. Viruses are composed of
   - O DNA
   - O RNA
   - O Either DNA or RNA
   - O Both DNA and RNA

4. Viruses are routinely cultured on agar medium.
   - O True
   - O False

5. Localized areas of cellular destruction and lysis caused by virus infection are called
   - O plaques
   - O spikes
   - O holes

6. The first viral vaccine was produced for
   - O rabies
   - O smallpox
   - O German measles

7. A virion is a complete virus particle that is consists of a protein capsid surrounding a nucleic acid molecule.
   - O True
   - O False

8. The lethal dose (LD50) of a virus is the dilution that contains a lethal dose large enough to destroy or damage
   - O all of the host cells or organisms
   - O half of the host cells or organisms
   - O one-quarter of the host cells or organisms

9. The infectious dose (ID50) of a virus is the dilution that contains an infectious dose large enough to destroy or damage
   - O all of the host cells or organisms
   - O half of the host cells or organisms
   - O one-quarter of the host cells or organisms

10. Virions range in size from 10 to 400 nanometers in diameter.
    - O True
    - O False

11. Most viruses are visible in the light microscope.
    - O True
    - O False

Chapter 3 Sample Test Questions on Procaryotic Cell Structure and Function

1. The cytoplasmic membrane performs which of the following functions?
   a. It retains the cytoplasm and its contents.
   b. It acts as a selectively permeable barrier, allowing some molecules to pass while preventing the movement of others.
   c. It provides the location for a variety of metabolic processes performed by the cell.
   d. It does all of these.

2. Which of the following is not true of bacterial plasmids?
   a. They can exist and replicate independently of the chromosome.
   b. They may carry genes for drug resistance.
   c. They are required for host growth and/or reproduction.
   d. They may carry genes that enhance the metabolic capabilities of the bacterium.

3. The lipopolysaccharide (LPS) that is found in the outer membrane of gram-negative bacteria is also known as
   a. exotoxin.
   b. teichoic acid.
   c. murein.
   d. endotoxin.

4. Gram-positive cells retain the primary stain whereas gram-negative cells do not because
   a. the alcohol removes the outer membrane and, therefore, the trapped stain from gram-negative bacteria.
   b. the alcohol shrinks the pores of the thick peptidoglycan layer of gram-positive cells, therefore, trapping the stain.
   c. Both of these are correct.
   d. Neither of these is correct.

5. Which of the following is not true of capsules?
   a. They help bacteria escape phagocytosis by host cells.
   b. They retain water and help prevent desiccation of the bacteria.
   c. They prevent entry of many bacterial viruses.
   d. All of these are true of capsules.

6. Chemotaxis implies that bacteria
   a. move toward an attractant.
   b. move away from a repellent.
   c. move toward an attractant and away from a repellent.
   d. neither move toward an attractant nor away from a repellent.

7. The term used to describe bacteria that have a round shape is
   a. coccus.
   b. bacillus.
   c. vibrio.
   d. coccobacillus.

8. The term used to describe bacteria that are rod shaped is
   a   coccus.
   b. bacillus.
   c. vibrio.
   d. coccobacillus.

9. The term used to describe bacteria that are shaped like curved rods is
   a. coccus.
   b. bacillus.
   c. vibrio.
   d. coccobacillus.

10. Bacteria that are shaped like rigid helices are
    a. vibrios.
    b. spirilla.
    c. spirochetes.
    d. coccobacilli.

11. Bacteria that are shaped like flexible helices are
    a. vibrios.
    b. spirilla.
    c. spirochetes.
    d. coccobacilli.

12. Square planar configurations that are formed when round bacteria remain attached to each other during reproduction are called
    a. streptococci.
    b. staphylococci.
    c. tetrads.
    d. sarcinae.

13. Cubic configurations that are formed when round bacteria remain attached to each other during reproduction are called
    a. streptococci.
    b. staphylococci.
    c. tetrads.
    d. sarcinae.

14. Proteins that are loosely attached to the cytoplasmic membrane are called __________ proteins.
    a. extrinsic
    b. intrinsic
    c. external
    d. internal

15. Proteins that are embedded within the cytoplasmic membrane and not easily extracted are called __________ proteins.
    a. extrinsic
    b. intrinsic
    c. external
    d. internal

16. Which of the following is not a function of the plasma membrane?
    a. It separates the cell from its environment.
    b. It serves as a selectively permeable barrier allowing some molecules to enter the cell while preventing others from doing so.
    c. It is the location of a variety of crucial metabolic processes including respiration, photosynthesis, lipid synthesis, and cell wall synthesis.
    d. All of these are functions of the plasma membrane.

17. Shrinkage of the plasma membrane away from the cell wall when the bacterium is placed in a hypertonic environment is called
    a. osmolysis.
    b. plasmolysis.
    c. hydrolysis.
    d. hypertonolysis.

18. Structurally asymmetric lipids with polar and nonpolar ends are said to be
    a. amphipathic.
    b. amphibolic.
    c. bilateral.
    d. None of these.

19. Penicillin inhibits cell wall synthesis, but cells will continue to grow normally in the presence of penicillin if they are maintained in a(n) __________ environment.
    a. hypotonic
    b. isotonic
    c. hypertonic
    d. nonpolar

20. Endospores represent a challenge to the fields of industrial and medical microbiology because
    a. they are resistant to harsh environments, thus allowing survival of sporing organisms under conditions in which nonsporing organisms would not survive.
    b. sporing organisms are often dangerous pathogens.
    c. Both of these answers are correct.
    d. Neither of these answers is correct.

21. The term used to describe bacteria that are oval shaped is:
    a. coccus.
    b. bacillus.
    c. vibrio.
    d. coccobacillus.

22. Mesosomes may
    a. play a role in cell wall formation during reproduction.
    b. aid in the process of chromosome replication and distribution.
    c. function during secretion of materials out of the cell.
    d. Do all of these.

23. T F   The cell membrane is a rigid structure that provides bacteria with their characteristic shapes.

24. T F   Cyanobacteria have membranous gas vacuoles that regulate their buoyancy but contain no lipid.

25. T F   Gram-positive bacteria have a thinner layer of peptidoglycan than gram-negative bacteria.

26. T F   Gram-positive bacteria have a structurally and chemically more complex cell wall than gram-negative bacteria.

27. T F   During the synthesis of flagella, the protein flagellin is added to the base of the flagellum so that the tip is constructed first and then pushed outward.

28. T F   The rotation of bacterial flagella is powered by ATP hydrolysis.

29. T F   Sedimentation coefficients are proportional to the molecular weight of a particle and are not affected by the volume and shape of the particle.

30. T F   Some scientists believe that mesosomes are artifacts generated during chemical fixation and therefore do not exist in living organisms.

31. T F   In the presence of attractants and repellents an organism will respond to the chemical with the highest molar concentration.

32. T F   Because mycoplasmas have no cell wall they must be maintained in an isotonic environment.

33. Granules of organic or inorganic material that are stockpiled by bacteria for future use are called __________.

34. Long multinucleated filaments are called __________.

35. The plasma membrane and everything contained within it is called a __________.

36. Branched networks of hyphae are called __________.

37. Enzymes that are secreted out of the cell to aid in the acquisition and digestion of environmental nutrients are called __________.

38. Molecules or regions of molecules that readily interact with water are said to be __________, whereas molecules or regions of molecules that are insoluble in water or do not readily interact with water are said to be __________.

39. A __________ is a well organized polysaccharide layer that lies outside the cell wall and is not easily removed.

40. A __________ layer consists of diffuse unorganized polysaccharide material that lies outside the cell wall and is easily removed

41. A __________ layer is a layer of protein or glycoprotein that exhibits a   pattern resembling floor tiles.

42. __________ is the process through which endospores are formed within a vegetative cell.

43. __________ is the process through which a vegetative cell is formed from an endospore.

44. Spirochetes move by means of a flexible filament, called a(n) __________, that is anchored to both ends of the organism.

45. Projections from the surface of a bacterium that are used to mediate its attachment to other bacteria during conjugation are called __________, whereas projections that mediate attachment to surfaces are called __________.

46. Bacteria have a region of the cytoplasm known as the __________, which is not bounded by a membrane but contains the chromosome.

47. The __________ space lies between the plasma membrane and the cell wall.

48. A small circular DNA molecule containing genes useful but not necessary to the bacterium is called a(n) __________.

49. The most widely accepted current model for membrane structures is called the __________ model.

50. T F In gram-negative bacteria, the space between the cell wall and theplasma membrane contains many proteins that participate in nutrient acquisition.

51. T F L forms may arise by the complete (or partial) loss of the cell wall. If the loss is complete, the L form will be stable because bacteria need a preexisting cell wall in order to construct new cell walls.

52. A bacterium travels in a straight line, called a __________, for a few seconds; then it stops and __________, moving in a different direction afterwards.

53. Grapelike clusters of round bacteria are referred to as __________.

54. Straight or branched chains of round bacteria are referred to as __________.

55. Bacteria that do not have a fixed shape are said to be __________.

56. Matching
1. monotrichous   a. A single flagellum
2. peritrichous   b. A single flagellum at each pole of an organism
3. lophotrichous   c. A cluster of flagella at one or both ends of an organism
4. amphitrichous   d. A relatively even distribution of flagella over the entire surface of the bacterium

57. T F Archaeobacterial cell walls like eubacterial cell walls contain peptidoglycan.

58. Helper proteins that aid in the proper folding of nascent polypeptide chains are called __________.
    a. scaffolding proteins
    b. molecular chapterones
    c. molecular folders
    d. folding proteins

59. Sites of direct contact or fusion between the outer membrane and the plasma membrane of gram-negative bacteria are called __________ sites.

60. Archaeobacterial membranes, like other membranes have a typical bilayer structure.

Chapter 4 Sample Test Questions on Eucaryotic Cell Structure and Function

1. Which of the following is useful in distinguishing between procaryotic and eucaryotic cells?
    a.   the use of DNA for the storage of genetic information
    b.   the presence of ribosomes for the synthesis of proteins
    c.   the presence of membrane-delimited organelles within the cytoplasm
    d.   All of these.

2. The endoplasmic reticulum is composed of flattened sacs called
    a.   thylakoids.
    b.   cristae.
    c.   cisternae.
    d.   vacuomes.

3. The Golgi apparatus is responsible for the synthesis of
    a.   ribosomes.
    b.   lysosomes.
    c.   mesosomes.
    d.   nucleosomes.

4. A(n) __________ is a special type of lysosome used by a cell to digest portions of its own cytoplasm.
    a.   suicide vacuole
    b.   turnover lysosome
    c.   recycling vacuole
    d.   autophagic vacuole

5. Which of the following is not a function of the mitochondrion?
    a.   tricarboxylic acid cycle enzyme reactions
    b.   electron transport
    c.   oxidative phosphorylation
    d.   All of these are functions of the mitochondrion.

6. The nucleolar organizer is a part of a specific chromosome that directs the synthesis of
    a.   transfer RNA (tRNA).
    b.   ribosomal RNA (rRNA).
    c.   messenger RNA (mRNA).
    d.   heterogeneous nuclear RNA (hnRNA).

7. A cell with a diploid number of 12 chromosomes undergoes mitosis. The number of chromosomes in each of the daughter cells is
    a.   24.
    b.   6.
    c.   12.
    d.   16.

8. A cell with a diploid number of 12 chromosomes undergoes meiosis. The number of chromosomes in each of the daughter cells is
   a. 24.
   b. 6.
   c. 12.
   d. 16.

9. Construction of flagella and/or cilia is directed by the
   a. axoneme.
   b. tubulin.
   c. centriole.
   d. basal body.

10. Intracellular structures that perform specific functions in cells are called
    a. organelles.
    b. organs.
    c. tissues.
    d. systems.

11. Protein fibers with a diameter of 4-7 nm that play a role in cell movement are called
    a. microtubules.
    b. microfilaments.
    c. intermediate filaments.
    d. mycelia.

12. Protein fibers with a diameter of 8-10 nm that help maintain cell shape and are involved in cell movement are called
    a. microtubules.
    b. microfilaments.
    c. intermediate filaments.
    d. mycelia.

13. Hollow fibers with a diameter of 25 nm that help maintain cell shape and are involved in cell movement are called
    a. microtubules.
    b. microfilaments.
    c. intermediate filaments.
    d. mycelia.

14. Endoplasmic reticulum (ER) with many ribosomes attached is called ___________ ER, whereas endoplasmic reticulum that is mostly devoid of ribosomes is called ___________ ER.

15. The process by which a cell imports large particles by enclosing them in vesicles pinched off from the plasma membrane is called
    a. exocytosis.
    b. phagocytosis.
    c. pinocytosis.
    d. nutrient cytosis.

16. The process by which a cell imports solutes by enclosing them in vesicles pinched off from the plasma membrane is called
    a. exocytosis.
    b. phagocytosis.
    c. pinocytosis.
    d. nutrient cytosis.

17. __________ are complexes consisting of several ribosomes attached to a single messenger RNA (mRNA) molecule.

18. Loosely organized genetic material containing those genes being actively expressed is called __________, whereas tightly organized genetic material containing those genes not being actively expressed is called __________.

19. An organism with two copies of each chromosome is said to be
    a. haploid.
    b. diploid.
    c. aneuploid.
    d. monoploid.

20. An organism with one copy of each chromosome is said to be
    a. haploid.
    b. diploid.
    c. aneuploid.
    d. monoploid.

21. Which of the following organelles is involved in the modification, packaging, and secretion of materials?
    a. lysosomes
    b. Golgi apparatus
    c. mitochondria
    d. centrioles

22. Enzymes needed to digest all types of macromolecules are found in
    a. lysosomes.
    b. Golgi apparatus.
    c. mitochondria.
    d. centrioles.

23. Proteins are synthesized on structures called
    a. mesosomes.
    b. lysosomes.
    c. ribosomes.
    d. chromosomes.

24. The major site of cell membrane synthesis is the
    a. nucleus.
    b. plasma membrane.
    c. lysosome.
    d. endoplasmic reticulum.

25. Which of the following parts of the Golgi complex receives materials from the ER?
    a. cis face
    b. trans face
    c. cisternal edges
    d. None of these are correct.

26. A cell with a diploid number of 12 chromosomes undergoes mitosis. The number of chromosomes in each daughter cell is:
    a. 24.
    b. 6.
    c. 8.
    d. 12.

27. A cell with a diploid number of 12 chromosomes undergoes meiosis. The number of chromosomes in each daughter cell is:
    a. 24.
    b. 6.
    c. 8.
    d. 12.

28. Long fibers containing microtubules and used for cellular locomotion are called __________, whereas short fibers containing microtubules and used for cellular locomotion are called

    __________.

29. The __________, a complex structure or set of structures lying underneath the plasma membrane, gives some cells their characteristic shape.

30. An arrangement of microtubules along the length of cilia and flagella is called the __________.
    a. axial bundle
    b. axial filament
    c. axoneme
    d. axillary tubule

31. The Golgi apparatus consists of vesicles called __________, which may be clustered in one region or scattered throughout the cell.
    a. cristae
    b. thylakoids
    c. cisternae
    d. vesiculae

32. The enzymes and electron carriers involved in electron transport and oxidative phosphorylation are located in the __________ membrane of the mitochondrion.
    a. inner
    b. outer
    c. inner and outer
    d. neither inner nor outer

33. The organelle responsible for processing light as an energy source is the
    a. kinetoplast.
    b. centriole.
    c. chloroplast.
    d. mitochondrion.

34. The formation of carbohydrate from carbon dioxide and water, the dark reaction of photosynthesis, takes place in the
    a. stroma.
    b. cytoplasm.
    c. thylakoid membrane.
    d. nucleus.

35. The trapping of light to form ATP and NADPH, the light reaction of photosynthesis, takes place in the
    a. stroma.
    b. cytoplasm.
    c. thylakoid membrane.
    d. nucleus.

36. The __________, a ring of granular and fibrous material surrounding the pores in a nuclear envelope, may facilitate the movement of material into and out of the nucleus.

37. The period between mitotic divisions during which most cell growth takes place is known as __________.

38. Cilia beat with two distinctive phases. With the __________ stroke, they stroke like an oar, thereby propelling the organism through the water. With the __________ stroke, cilia bend along their length while they are pulled forward.

39. Lateral hairs on a flagellum are called __________ __________.

40. T F New ER is produced through expansion of existing ER.

41. T F Mitosis is the process in eucaryotes of cytoplasmic division and nuclear division with the distribution of chromosomes.

42. T F Chloroplasts are the site for both the light and dark reactions of photosynthesis.

43. T F The majority of mitochondrial proteins are manufactured under the direction of the nuclear DNA by cytoplasmic ribosomes.

44. T F DNA is replicated prior to the start of mitosis.

45. T F The membrane bound structure that houses the chromosomes of eucaryotes is called the nucleus.

46. T F Eucaryotes usually have more than one chromosome for the storage of genetic information.

47. T F The endoplasmic reticulum is mostly devoid of ribosomes (smooth) when cells are producing large quantities of lipid.

48. T F Although similar in function, the eucaryotic ribosome is generally larger and more complex than the procaryotic ribosome.

49. In filaments in which there are no cross-walls between adjacent cells, the cells are said to be
    a. monokaryotic.
    b. multinucleated.
    c. coenocytic.
    d. None of these.

50. T F Coenocytic cells occur when mitosis takes place in the absence of cytokinesis.

51. T F Like procaryotic cells, most eucaryotic cells have an external cell wall.

52. The inner membrane system of chloroplasts consists of flattened sacs called __________ that form stacks called __________.

53. T F The nucleolus is a highly active region of the chromatin that is involved in the synthesis of ribosomes.

54. T F The major difference between the storage vacuoles of eucaryotes and the inclusion bodies of procaryotes is that the former are membrane bound whereas the latter are not.

55. T F Electron transport reactions similar to those that occur in the mitochondria of eucaryotes also occur in procaryotes but take place on the plasma membrane in the latter cells.

56. T F In addition to being more complex chemically, the basic structure of the plasma membrane of eucaryotic cells is very different from that of procaryotic cells.

57. Stacks of cisternae in the Golgi apparatus are referred to as:
    a. stigmata
    b. golgisomes
    c. dictyosomes
    d. lamellisomes

Chapter 16 Sample test Questions on The Viruses: Introduction and General Characteristics

1. Who of the following demonstrated that the causative agent of tobacco mosaic disease would pass through filters designed to retain bacteria but thought that the agent was a nonreproducing toxin?
    a. Jenner
    b. Ivanowski
    c. Beijerinck
    d. Pasteur

2. Who of the following showed that the filterable agent that caused tobacco mosaic disease was capable of reproduction?
    a. Jenner
    b. Ivanowski
    c. Beijerinck

3. Who of the following devised a method for enumerating viruses that could infect bacteria?
    a. Twort
    b. d'Herelle
    c. both Twort and d'Herelle
    d. neither Twort nor d'Herelle

4. Who of the following separated tobacco mosaic virus particles into protein and nucleic acid?
    a. Bawden
    b. Pirie
    c. both Bawden and Pirie
    d. neither Bawden nor Pirie

5. Which of the following is not true of viruses?
    a. Viruses are acellular.
    b. Viruses consist of one or more molecules of DNA or RNA enclosed in a coat of protein and sometimes in other more complex layers.
    c. Viruses are obligate intracellular parasites.
    d. All of these are true of viruses.

6. Viruses contain
    a. DNA.
    b. RNA.
    c. either DNA or RNA but not both simultaneously.
    d. both DNA and RNA simultaneously.

7. Animal viruses can be cultivated in
    a. suitable host animals.
    b. embryonated eggs.
    c. tissue cultures (monolayers of animal cells).
    d. All of these.

8. In which phase of the bacterial growth cycle is the cultivation of bacterial viruses most efficient?
    a. lag
    b. log
    c. stationary
    d. decline

9. Which of the following cannot be used to cultivate plant viruses?
    a. cultures of separated plant cells
    b. whole plants
    c. plant protoplast cultures
    d. All of these can be used to cultivate plant viruses.

10. Which of the following is more sensitive to small differences in size among various viruses?
    a. differential centrifugation
    b. gradient centrifugation
    c. Both are sensitive to small differences in size.
    d. Neither is sensitive to small differences in size.

11. Which of the following cannot be used to denature and precipitate contaminants from viral preparations?
    a. heat
    b. pH changes
    c. organic solvents
    d. All of these can be used to denature and precipitate contaminants.

12. T F   Enzymatic degradation of cellular proteins and/or nucleic acids cannot be used in the purification of viruses because the viruses are usually more sensitive to the actions of these enzymes than the contaminating host cell materials.

13. Which of the following is most useful in determining the viability of a viral preparation?
    a. a direct (electron microscopic) count
    b. hemagglutination
    c. a plaque assay
    d. All are equally useful in determining the viability of a viral preparation.

14. Virion size ranges from
    a. 10-40 nm.
    b. 10-400 nm.
    c. 100-400 nm.
    d. 100-4,000 nm.

15. Which of the following is most true of a viral DNA genome?
    a. It uses the same four nitrogenous bases found in procaryotic and eucaryotic DNA.
    b. It may have the normal bases found in procaryotic and eucaryotic DNA, or it may have one or more unusual bases.
    c. It is usually composed of unusual bases unique to viruses.
    d. Viruses do not use DNA for their genome.

16. The protein coat surrounding the viral genome is called the
    a. capsule.
    b. capsid.
    c. matrix.
    d. envelope.

17. The function of the viral protein coat is to
    a. protect the viral genetic material.
    b. aid in the transfer of the viral genetic material between host cells.
    c. both protect the viral genetic material and aid in the transfer of the viral genetic material between host cells.
    d. neither protect the viral genetic material nor aid in the transfer of the viral genetic material between host cells.

18. The two major types of symmetry found in viruses are
    a. icosahedral and radial.
    b. icosahedral and helical.
    c. helical and radial.
    d. radial and bilateral.

19. Viral envelopes are composed of
    a. proteins.
    b. lipids.
    c. carbohydrates.
    d. All of these.

20. Which of the following is true about viral envelopes?
    a. The envelope proteins are virus specific.
    b. The envelope lipids and carbohydrates are derived from the host.
    c. Both of these answers are true.
    d. Neither of these answers is true.

21. Enzymes found in virus particles
    a   may be required for viral attachment to host cells.
    b. may be involved in viral entry into host cells.
    c. may be involved in the replication of viral nucleic acid.
    d. All of these.

22. Viruses are subgrouped according to
    a. nucleic acid type (DNA or RNA).
    b. nucleic acid strandedness (single or double).
    c. presence or absence of an envelope.
    d. All of these.

23. Who of the following instituted vaccination procedures as a preventive measure against smallpox?
    a. Pasteur
    b. Jenner
    c. Lister
    d. Lwoff

24. A complete virus particle is called a
    a. capsid.
    b. nucleocapsid.
    c. virion.
    d. cell.

25. Which of the following is not true about the word Virus?
    a. It is of Latin origin meaning poison.
    b. It was used to describe disease-causing agents long before such agents were observed.
    c. It is currently used to describe small acellular entities that cause disease.
    d. All of these are true about the word Virus.

26. Centrifugation of a suspension at various speeds in order to separate particles of various sizes is called __________ centrifugation.
    a. isopycnic
    b. density
    c. differential
    d. variable

27. In gradient density centrifugation, which of the following is true?
    a. Particles will continue to settle toward the bottom of the tube for as long as the centrifugation is continued.
    b. Particles will come to rest when the density of the surrounding medium is equal to the density of the particle even if the centrifugation is continued longer.
    c. The smallest particle will sediment fastest.
    d. Large particles will sediment farther than small particles.

28. Viruses with single-stranded RNA as their genome for which the base sequence is the same as the viral mRNA are said to be ___________ viruses.
    a. plus-stranded
    b. minus-stranded
    c. mRNA-like
    d. None of these.

29. A ___________ genome exists as several separate, nonidentical molecules that may be packaged together or separately.
    a. diploid
    b. segmented
    c. polyploid
    d. fractionated

30. Which of the following is true of viruses in the extracellular phase?
    a. They possess many different enzymes.
    b. They can reproduce independently of living cells.
    c. Both of these answers are true.
    d. Neither of these answers is true.

31. Viral capsids are generally constructed without any outside aid once the subunits have been synthesized. This process is called
    a. spontaneous assembly.
    b. self-assembly.
    c. spontaneous maturation.
    d. self-maturation.

32. Glycoprotein spikes protruding from the outer surface of viral envelopes may be involved in
    a. viral recognition of host cells.
    b. viral attachment to host cells.
    c. both viral recognition of and viral attachment to host cells.
    d. neither viral recognition of nor viral attachment to host cells.

33. Some complex viruses have icosahedral symmetry in the head region and helical symmetry in the tail. Overall, these viruses are said to have ___________ symmetry.
    a. bilateral
    b. binal
    c. complex
    d. None of these.

34. In a plaque assay, each viral plaque is assumed to have arisen from the reproduction of __________ virion(s).
    a. one
    b. two
    c. one or more
    d. more than one

35. In a plaque assay, the number of infectious virions is usually __________ the number of virus particles present.
    a. equal to
    b. less than
    c. greater than
    d. not predictable relative to

36. Who of the following demonstrated that leukemia in chickens was caused by a filterable agent?
    a. Ellerman
    b. Bang
    c. both Ellerman and Bang
    d. neither Ellerman nor Bang

37. In an enveloped virus, the part of the virus including the nucleic acid genome and the surrounding protein coat but not the envelope is called the
    a. capsid.
    b. nucleocapsid.
    c. matrix.
    d. virion.

38. Hemagglutination is
    a. the clumping together of red blood cells in the presence of a viral suspension.
    b. the binding of iron in the process of a viral suspension.
    c. the clumping together of infected cells in the presence of a viral suspension.
    d. None of these.

39. Who showed that hoof and mouth disease was caused by a filterable virus?
    a. Loeffler and Frosch
    b. Ellerman and Bang
    c. Bawden and Pirie
    d. None of these.

40. Viral capsid protein subunits are called
    a. auxomers.
    b. elastomers.
    c. protomers.
    d. viromers.

41. T F   Like bacteria and eucaryotic microorganisms, viruses can be cultured on artificial media.

42. T F   All capsids, regardless of the type of symmetry, tend to be rigid.

43. T F   Viruses in the extracellular state possess few, if any, active enzymes.

44.T F   Viruses have typical cellular structure like other living organisms.

45.T F   Bacterial viruses are so named because they have the same procaryotic cell structure as their bacterial hosts.

46.T F   The primary categorization of viruses is on the basis of host preference.

47.T F   The presence or absence of an envelope is not useful in classifying viruses because any given virus may at one time have an envelope and at another time not have an envelope.

48.T F   In viral classification, greatest weight is given to the intracellular location of viral replication.

49.T F   The largest of the viruses are larger than some small bacteria and are large enough to be seen with a light microscope.

50.   Viruses that infect bacteria were first discovered by __________ but he   did not follow up on the discovery.

51.   Tobacco mosaic virus was crystallized by __________ and shown to be   mostly protein.

52.   Viruses that are polyhedrons with 20 sides are said to have __________ symmetry.

53.   Icosahedral viruses are constructed from ring- or knob-shaped units called __________.

54.   __________ are glycoprotein spikes protruding from the outer surface of the viral envelope.

55.   The culture of cells in a plaque assay is usually overlayed with   __________ in order to limit the spread of the virus so that only the cells immediately adjacent to the infected cells will be infected by the newly produced virions, thereby forming localized areas of cellular destruction (plaques).

56.   __________ demonstrated that the causative agent of yellow fever in humans was filterable and could be transmitted by mosquitoes.

57.   Microscopic or macroscopic degenerative changes or abnormalities in infected host cells and tissues are called __________.

58.   Localized areas of destruction occurring on plants that have been infected by a virus are referred to as __________ lesions.

59.   Which of the following are usually sensitive to ether treatment?
    a.   enveloped viruses
    b.   nonenveloped viruses
    c.   both enveloped and nonenveloped viruses
    d.   neither enveloped nor nonenveloped viruses

60.T F   Viruses have been somewhat difficult to study because they cannot be cultured on artificial media.

Answers to Sample Test Questions from Video 2

1.	Yes	11.	E
2.	False	12.	False
3.	E	13.	Envelope
4.	False	14.	False
5.	False	15.	True
6.	True	16.	DNA
7.	C	17.	True
8.	Flagellum	18.	True
9.	False	19.	True
10.	False	20.	Metabolism

Answers to Sample Test Questions from Text Chapter 3 (Procaryotic Cell Structure and Function) Web Site

1.	4	6.	2
2.	1	7.	1
3.	1	8.	4
4.	1	9.	3
5.	2	10.	1

Answers to Sample Test Questions from Text Chapter 4 (Eucaryotic Cell Structure and Function) Web Site

1.	1	6.	2
2.	1	7.	1
3.	1	8.	2
4.	1	9.	2
5.	1	10.	3

Answers to Sample Test Questions from Text Chapter 16 (The Viruses: Introduction and Characteristics) Web Site

1.	1	7.	1
2.	2	8.	2
3.	3	9.	2
4.	2	10.	1
5.	1	11.	2
6.	2		

Answers to Chapter 3 Sample Test Questions on Procaryotic Cell Structure and Function

The questions in this test bank have been rated for difficulty. The ratings run from 1 (for the easiest) to 8 (for the most difficult)

#	Answer/Diff.		#	Answer/Diff.
	D / 4		30	T / 5
2	C / 6		31	F / 8
3	D / 4		32	F/8
4	C / 7		33	inclusion bodies/7
5	D / 7		34	hyphae / 5
6	C / 5		35	protoplast / 7
7	A / 1		36	mycelia / 5
8	B / 1		37	exoenzymes / 5
9	C / 1		38	hydrophilic; hydrophobic / 5
10	B / 3		39	capsule / 5
11	C / 3		40	slime / 5
12	C / 4		41	structured / 5
13	D / 4		42	Sporulation / 4
14	A / 5		43	Germination / 4
15	B / 5		44	axial filament / 5
16	D / 6		45	pili; fimbriae / 6
17	B / 7		46	nucleoid / 4
18	A / 6		47	periplasmic / 5
19	B / 8		48	plasmid / 5
20	C / 8		49	fluid mosaic / 6
21	D / 2		50	T / 7
22	D / 6		51	T / 7
23	F / 5		52	run; tumbles / 5
24	T / 7		53	staphylococci / 2
25	F / 5		54	streptococci / 2
26	F / 5		55	pleomorphic / 5
27	F / 7		56	1 A, 2 D, 3 C, 4 B / 5
28	F / 8		57	F / 7
29	F / 8		58	B / 5
30	T / 5		59.	adhesion/ 4
			60.	F / 6

Answers to Chapter 4 Sample Test Questions on Eucaryotic Cell Structure and Function

The questions in this test bank have been rated for difficulty. The ratings run from 1 (for the easiest) to 8 (for the most difficult)

#	Answer/Diff.	#	Answer/Diff.
1	C / 5	32	A / 4
2	C / 7	33	C / 2
3	B / 8	34	A / 4
4	D / 8	35	C / 4
5	D / 5	36	annulus / 8
6	B / 8	37	interphase / 6
7	C / 5	38	effective; recovery / 7
8	B / 5	39	flimmer filaments / 8
9	D / 8	40	T / 5
10	A / 3	41	F / 7
11	B / 7	42	T / 4
12	C / 7	43	T / 6
13	A / 6	44	T / 6
14	rough; smooth / 5	45	T / 4
15	B / 6	46	T / 5
16	C / 6	47	T / 7
17	Polysomes / 8	48	T / 3
18	euchromatin; heterochromatin /	49	C / 5
19	B / 4	50	T / 7
20	A / 5	51	F / 3
21	B / 5	52	thylakoids; grana / 5
22	A / 5	53	T / 5
23	C / 1	54	T / 5
24	D / 4	55	T / 5
25	A / 6	56	F / 5
26	D / 4	57.	C / 6
27	B / 4		
28	flagella; cilia / 4		
29	pellicle / 8		
30	C / 7		
31	C / 5		

Answers to Chapter 16 Sample Test Questions on The Viruses: Introduction and General Considerations

The questions in this test bank have been rated for difficulty. The ratings run from 1 (for the easiest) to 8 (for the most difficult).

#	Answer/Diff.	#	Answer/Diff.
1	B / 6	31	B / 5
2	C / 6	32	C / 2
3	B / 7	33	B / 7
4	C / 5	34	A / 4
5	D / 4	35	B / 5
6	C / 5	36	C / 5
7	D / 3	37	B / 6
8	B / 5	38	A / 5
9	D / 6	39	A / 6
10	B / 5	40	C / 6
11	D / 6	41	F / 3
12	F / 8	42	F / 6
13	C / 8	43	T / 5
14	B / 6	44	F / 6
15	B / 7	45	F / 2
16	B / 4	46	T / 1
17	C / 3	47	F / 8
18	B / 5	48	F / 4
19	D / 5	49	T / 5
20	C / 8	50	Twort / 8
21	D / 5	51	Stanley / 7
22	D / 5	52	icosahedral / 6
23	B / 4	53	capsomeres / 7
24	C / 5	54	Peplomeres / 8
25	D / 7	55	agar / 5
26	C / 5	56	Walter Reed / 7
27	B / 7	57	cytopathic effects / 6
28	A / 6	58	necrotic / 5
29	B / 5	59	A / 4
30	D / 4	60	T / 5

Microbial Cell Biology
Program 3
Metabolism

It is in the fueling reactions that bacteria display their extraordinary metabolic diversity and versatility. Bacteria have evolved to thrive in almost all natural environments, regardless of the nature of available sources of carbon, energy, and reducing power....The collective metabolic capacities of bacteria allow them to metabolize virtually every organic compound on this planet....

–F. C. Neidhardt
J. L. Ingraham, and M. Schaechter

One of microbiology's fifty most significant events in the past 125 years:

1878
Joseph Lister publisherd his study of lactic acid fermentation of milk, demonstrating the specific cause of milk souring. His research was conducted using the first method developed for isolating a pure culture of a bacterium, which he named *Bacterium lactis*.

Student Assignments

- **Watch** Video 3: Metabolism
- **Read**   Chapter 9: Metabolism: The Generation of Energy in *Microbiology*, 4th ed., Prescott/Harley/Klein, McGraw-Hill
- **Lab Experiments** from *Laboratory Exercises in Microbiology*, 4th ed., Harley/Prescott, McGraw-Hill
  - Exercise 22: Carbohydrates I: Fermentation and β-Galactosidase Activity
  - Exercise 50: Determination of a Bacterial Growth Curve: Classical and Two-hour Methods

Lesson Overview

Video 3: Metabolism
The metabolic pathways that produce energy create important environmental transformations. Although living organisms have diverse ways of meeting their energy needs, there is an amazing similarity between all life forms as they carry out metabolism directed to the construction and use of biological molecules.

Chapter 9: Metabolism: The Generation of Energy
This chapter presents an overview of metabolism beginning with carbohydrate degradation and the aerobic generation of ATP through electron transport. Fermentation and anaerobic respiration are examined, followed by the catabolism of lipids, proteins, and amino acids. The chapter concludes with discussions of the function of inorganic molecules as electron acceptors and the trapping of energy by photosynthesis.

Learning Objectives

After successfully completing this unit, the student will be able to:

- discuss the difference between catabolism and anabolism
- describe the various pathways for the catabolism of glucose to pyruvate
- discuss the tricarboxylic acid (TCA) cycle and its central role in aerobic metabolism
- describe the electron transport process, and compare and contrast the electron transport system of eucaryotes with that of procaryotes
- contrast the two major proposed mechanisms for oxidative phosphorylation (i.e., the chemiosmotic hypothesis and the conformational change hypothesis)
- discuss the different electron acceptors used during aerobic respiration, fermentation, and anaerobic respiration
- describe in general terms the catabolism of molecules other than carbohydrates
- discuss the photosynthetic light reactions, and compare and contrast the light reactions of eucaryotes (and cyanobacteria) with those of green (or purple) photosynthetic bacteria

Course Integrator

On Overview of Metabolism (p. 163)
Transparency　　　　　　　　　Three Stages of Catabolism Figure 9.1
Visual Resource Library　　　　Three Stages of Catabolism Figure 9.1
Microbes in Motion II　Microbial Metabolism & Growth/Microbial Metabolism
Video *Unseen Life on Earth*　Unit #3 Metabolism

The Breakdown of Glucose to Pyruvate (p. 165)

Transparency	Glycolysis Figure 9.3
Visual Resource Library	Glycolysis Figure 9.3
Transparency	Pentose Phosphate Pathway Figure 9.4
Visual Resource Library	Pentose Phosphate Pathway Figure 9.4
Visual Resource Library	Transketolase & Transaldolase Figure 9.5
Transparency	Entner-Doudoroff Pathway Figure 9.6
Visual Resource Library	Entner-Doudoroff Pathway Figure 9.6
Visual Resource Library	Glycolysis Figure AII.1
Visual Resource Library	Pentose Phosphate Pathway Figure AII.2

The Tricarboxylic Acid Cycle (p. 169)

Transparency	Tricarboxylic Acid Cycle Figure 9.7
Visual Resource Library	Tricarboxylic Acid Cycle Figure 9.7
Visual Resource Library	Tricarboxylic Acid Cycle Figure AII.4

Electron Transport and Oxidative Phosphorylation (p. 169)

Transparency	Respiratory Chain of *E coli* Figure 9.9
Visual Resource Library	Respiratory Chain of *E. coli* Figure 9.9
Transparency	Chemiosmotic Hypothesis Figure 9.10
Visual Resource Library	Chemiosmotic Hypothesis Figure 9.10
Transparency	Mitochondrial Transport Chain Figure 9.11
Visual Resource Library	Mitochondrial Transport Chain Figure 9.11
Visual Resource Library	ATP synthesis Figure 9.12
Microbes in Motion II	Bacterial Structure & Function/Cell Membrane /Energy Production /Electron Transport Chain

Fermentations (p. 174)

Transparency	NADH Oxidation in Fermentation Figure 9.13
Visual Resource Library	NADH Oxidation in Fermentation Figure 9.13
Transparency	Microbial Fermentations Figure 9.14
Visual Resource Library	Microbial Fermentations Figure 9.14

Microbial Terms Used in Video 3 in Order of Appearance with Page Indexing to the Textbook (*Microbiology*, 4th ed, Prescott/Harley/Klein, McGraw-Hill)

energy 477	chemolithotrophys 180	activated sludge 869
metabolism 163	fermentation 174	anaerobic digester 430
ATP 152	primary producers 832	methanogens 871
anabolism 190	consumers 832	carbon fixation 837
catabolism 164	ecosystems 832	
photosynthesis 182		

tube worms   568
S. cerevisiae   524
wort   925
growth curve   114

Web Links Related to Video 3 (Metabolism) and Readings in Chapter 9 (Metabolism: The Generation of Energy)(Go to <u>www.mhhe.com/prescott</u>, Student on Line learning Center, Double Click on Chapter 9, Double Click on Web Sites)

Simplified Diagram of Cellular Metabolism
<u>http://www.accessexcellence.org/ab/gg/cellmetab.html</u>

Graphics Gallery (Scroll down to "Cell processes")   The Graphics Gallary at the Acess Excellence Web site is a series of labeled diagrams with explanations representing the most important processes of biotechnology.
<u>http://www.accessexcellence.org/ab/gg</u>

Enzyme Biochemistry Chapter MIT Biology Hypertextbook
<u>http://esg-www.mit.edu8001/esgbio/eb/ebdir.html</u>

Biochemistry and Molecular Biology (Biosciences).   A catalog of useful Web sites.
<u>http://mcb.harvard.edu/biolinks.html</u>

Microbial Terms Used in Chapter 9 (Metabolism: The Generation of Energy)

accessory pigments 182
acetyl coenzyme A   169
adenosine 5'-phospho-
    sulfate   180
alcoholic fermentation
    175
amphibolic pathways
165
anabolism   164
anaerobic respiration
176
anoxygenic
    photosynthesis   185
antenna   182
ATP synthase   173

bacteriochlorophyll 185
β-oxidation   178
butanediol fermentation
    175
carotenoids   182
catabolism   164
chemiosmotic
    hypothesis   171
chemolithotroph   180
chlorophylls   182
citric acid cycle   169
dark reactions   184
deamination   176
denitrification   176
electron transport chain   170

Embden-Meyerhof
    pathway   166
Enter-Doudoroff
    pathway   168
fermentation   174
glycolysis   166
glycolytic pathway
    166
heterolytic
fermenters
    175
hexose monophos-
    phate pathway   167

Sample Test Questions from Video 3 (Metabolism)

1. Different organisms have different energy needs.
   a. True
   b. False

2. An organisms particular __________ determines which energy producing strategy is successful and therefore which organism will survive.

3. The one molecule shared by all living organisms is __________.

4. The flow of matter and energy is termed __________.

5. Cells produce energy in the form of __________ and use that energy to drive their chemical reactions.

6. Converting light energy in to ATP is called __________.

7. Converting inorganic chemical energy into ATP is called __________.

8. Converting organic chemical energy into ATP is
   a. fermentation
   b. respiration
   c. Both a and b
   d. photosynthesis

9. Specialized proteins that catalyze metabolism are called __________.

10. Microorganisms that use the energy from the sun to create carbon rich compounds are called __________ __________.

11. Deep beneath the surface of the ocean there are organisms that use __________ instead of sunlight as the basis of their energy metabolism. These organisms are called __________.

12. One of the oldest microbial processes is
    a. fermentation.
    b. respiration.
    c. photosynthesis.
    d. All of the above.

Sample Test Questions from Text Chapter 9 (Metabolism: The Generation of Energy) Web Site

1. The sum total of all the reactions occurring in the cell is called
   o catabolism
   o anabolism
   o metabolism

2. The synthesis of complex molecules from simpler ones with the input of energy is called
   o catabolism
   o anabolism
   o metabolism

3. The process by which large, complex molecules are broken down into smaller, simpler molecules, with the release of energy is called
   o catabolism
   o anabolism
   o metabolism

4. In heterotrophic microorganisms, carbohydrates and other nutrients
   o are oxidized to release energy
   o supply carbon or other building blocks for the synthesis of new molecules
   o all of the above

5. The glycolytic pathway (glycolysis)
   o is found in all major groups of microorganisms
   o is called the Krebs cycle
   o operates only in the presence of oxygen
   o all of the above

6. ATP molecules are generated during oxidative phosphorylation.
   o True
   o False

7. The energy-yielding process in which the electron transport chain acceptor is an inorganic molecule other than oxygen is called
   o oxidative phosphorylation
   O fermentation
   O anaerobic respiration

8. Nitrate reduction refers to the process by which nitrate is used as the electron acceptor at the end of the electron transport chain
   O True
   O False

9. Denitrification is a process by which ammonia in the soil is oxidized to nitrate.
   O True
   O False

10. Cyanobacteria are photosynthetic organisms.
    O True
    O False

Chapter 9 Sample Test Questions on Metabolism: The Generation of Energy

1. Which of the following is not one of the three stages of metabolism described by Lehninger?
   a. hydrolysis of large macromolecules (proteins, polysaccharides, and lipids)
   b. tricarboxylic acid cycle
   c. photosynthesis
   d. degradation of amino acids, fatty acids, glycerol, and monosaccharides to simpler molecules

2. Carbohydrates and other nutrients are used as cellular constituents
   a. with modification.
   b. without modification.
   c. both with and without modification.
   d. neither with nor without modification.

3. The synthesis of ATP from ADP and Pi, when coupled with the exergonic enzymatic breakdown of a high-energy molecule, is called __________ phosphorylation.
   a. chemiosmotic
   b. oxidative
   c. substrate-level
   d. conformational change

4. The pentose phosphate pathway provides
   a. ATP.
   b. NADPH.
   c. four- and five-carbon sugars for amino acid and nucleic acid synthesis.
   d. All of these.

5. How many molecules of ATP are needed to reduce one molecule of carbon dioxide to carbohydrate?
   a. 2
   b. 3
   c. 4
   d. 10-12

6. How many molecules of NADPH are needed to reduce one molecule of carbon dioxide to carbohydrate?
   a. 2
   b. 3
   c. 4
   d. 10-12

7. How many quanta of light are needed to reduce one molecule of carbon dioxide to carbohydrate?
   a. 2
   b. 3
   c. 4
   d. 10-12

8. The major function(s) of the TCA cycle is(are)
   a. energy production.
   b. provision of carbon skeletons for biosynthesis of cell components.
   c. both energy production and provision of carbon skeletons for biosynthesis of cell components.
   d. neither energy production nor provision of carbon skeletons for biosynthesis of cell components.

9. The number of ATP molecules generated when electrons are passed from NADH or reduced FAD (FADH) to oxygen is called the
   a. utilization ratio.
   b. energy yield.
   c. P/O ratio.
   d. phosphorylation coefficient.

10. Most aerobically generated ATP comes from
    a. glycolysis.
    b. the TCA cycle.
    c. the electron transport chain.
    d. All of these generate equal amounts of ATP.

11. T F Oxygen is required for the regeneration of NAD from NADH.

12. Mitochondrial electron transport takes place
    a. on the outer mitochondrial membrane.
    b. on the inner mitochondrial membrane.
    c. in the fluid matrix of the mitochondria.
    d. both on the inner and on the outer mitochondrial membrane.

13. Photosynthetic electron transport takes place
    a. on the thylakoid membranes.
    b. in the grana.
    c. in the stroma.
    d. both on the thylakoid membranes and in the grana.

14. The sum total of all chemical reactions occurring in a cell is called
    a. anabolism.
    b. catabolism.
    c. metabolism.
    d. None of these.

15. The breakdown of larger, more complex molecules into smaller, simpler ones with the release
    and trapping of some energy contained within those molecules is called
    a. anabolism.
    b. catabolism.
    c. metabolism.
    d. None of these.

16. The synthesis of complex molecules from simpler ones with the input of energy is called
    a. anabolism.
    b. catabolism.
    c. metabolism.
    d. None of these.

17. ___________ pathways are those that function both catabolically and anabolically.

18. Production of ATP using the energy liberated by the electron transport system is generally
    referred to as ___________ phosphorylation.
    a. chemiosmotic
    b. oxidative
    c. substrate-level
    d. conformational change

19. Two hypotheses have been proposed for the production of ATP from the electron transport
    system. They are the ___________ and the ___________ hypotheses.

20. When cells switch from anaerobic to aerobic conditions, a drastic reduction in the rate of sugar
    catabolism, called the ___________ occurs.

21.Oxygen is used as the terminal electron acceptor for oxidizing NADH to NAD in __________ respiration, but molecules other than oxygen are used as the terminal electron acceptors in __________ respiration.

22.The process by which electrons are excited from a chlorophyll molecule then returned to that molecule to drive the synthesis of ATP is most specifically referred to as
a. photosynthesis.
b. cyclic photophosphorylation.
c. acyclic photophosphorylation.
d. noncyclic photophosphorylation.

23.The process by which electrons excited from a chlorophyll molecule flow from water to NADPH to drive the synthesis of ATP is most specifically referred to as
a. photosynthesis.
b. cyclic photophosphorylation.
c. acyclic photophosphorylation.
d. noncyclic photophosphorylation.

24.The net gain of ATP per molecule of glucose metabolized aerobically in eucaryotes is
a. 2.
b. 36.
c. 38.
d. 85.

25.The net gain of ATP per molecule of glucose metabolized aerobically in procaryotes is
a. 2.
b. 36.
c. 38.
d. 85.

26.The net gain of ATP per molecule of glucose metabolized anaerobically in eucaryotes is
a. 2.
b. 36.
c. 38.
d. 85.

27.The net gain of ATP per molecule of glucose metabolized anaerobically in procaryotes is
a. 2.
b. 36.
c. 38.
d. 85.

28. In the TCA cycle, two carbons in the form of ___________ are added to oxaloacetate at the start of the cycle.
    a. acetyl-CoA
    b. ethanol
    c. carbon dioxide
    d. methanol

29. In the TCA cycle, two carbons are removed from citric acid in the form of ___________, thereby regenerating oxaloacetate to complete the cycle.
    a. acetyl-CoA
    b. ethanol
    c. carbon dioxide
    d. methanol

30. The mitochondrial electron transport system is arranged into four complexes of carriers connected to each other by ___________ and ___________.

31. In aerobic conditions, it takes ___________ sugar to produce the same amount of ATP when compared to anaerobic conditions.
    a. more
    b. less
    c. the same amount of

32. Organisms that directly reduce almost all of their pyruvate to lactate are called ___________ fermenters.

33. Organisms that form both lactate and substantial amounts of other products are called ___________ fermenters.

34. Which of the following can be used as electron acceptors during anaerobic respiration?
    a. nitrate
    b. sulfate
    c. carbon dioxide
    d. All of these.

35. Disaccharides and polysaccharides can be processed as nutrients after first being cleaved to monosaccharides by either ___________ or ___________.

36. Fatty acids are metabolized by the ___________ pathway.
    a. alpha-oxidation
    b. beta-oxidation
    c. gamma-oxidation
    d. delta-oxidation

37. During breakdown of fatty acids, carbons are removed __________ at a time as __________.
    a. one; carbon dioxide
    b. one; methane
    c. two; acetyl-CoA
    d. two; ethanol

38. Proteins are catabolized by hydrolytic cleavage to amino acids by the action of enzymes called __________.

39. Amino acids are processed first by the removal of the amino group through
    a. deamination.
    b. transamination.
    c. either deamination or transamination.
    d. neither deamination nor transamination.

40. In addition to their place in various food chains, photosynthetic organisms are responsible for replenishing our supply of __________.

41. Which of the following statements is correct?
    a. Procaryotes usually have a higher P/O ratio than eucaryotes.
    b. Procaryotes usually have a lower P/O ratio than eucaryotes.
    c. Procaryotes usually have the same P/O ratio as eucaryotes.

42. The hypothesis that proton motive force drives ATP synthesis is called the __________ hypothesis.

43. The hypothesis that the energy released during electron transport causes a change in the shapes of enzymes and thereby causes a change in their binding affinities is called the __________ __________ hypothesis.

44. Which of the following may be used as sources of energy by chemolithotrophs?
    a. hydrogen gas
    b. nitrogen
    c. sulfur
    d. All of these.

45. T F All eucaryotic organisms use the same terminal oxidase for the electron transport system, whereas procaryotes use several different terminal oxidases.

46. T F Although the main purpose of fermentation reactions is the oxidation of NADH to NAD, some also generate additional ATP for the organism.

47. T F An organism may use glycolysis and the pentose phosphate pathway simultaneously.

48. T F Uncouplers stop ATP synthesis without inhibiting electron transport.

49.T F Although most metabolic reactions are freely reversible, a few irreversible catabolic steps are bypassed during biosynthesis with special enzymes that catalyze the reverse reaction in order to allow for independent regulation of catabolic and anabolic pathways.

50.T F Photosynthetic organisms serve as the base for most food chains in the biosphere.

51.T F The electron transport systems in eucaryotes and procaryotes use different electron carriers.

52.T F All eucaryotic organisms use the same terminal oxidase for their electron transport systems.

53.T F Different procaryotic organisms use different terminal oxidases for their electron transport systems.

54.T F Metabolic pathways can be either catabolic or anabolic but not both.

55. Which of the following best describes the evidence that currently exists about oxidation phosphorylation?
   a. The evidence currently supports the chemiosmotic hypothesis.
   b. The evidence currently supports the conformation change hypothesis.
   c. The evidence indicates that aspects of both hypotheses may be correct.
   d. The evidence favors neither the chemiosmotic nor the conformational change hypothesis but rather it suggests the existence of a different and as yet unknown mechanism.

56.T F Procaryotic and eucaryotic electron transport chains differ not only in construction but also in the fundamental principles upon which they operate.

57.T F All eucaryotic photosynthesizers are oxygenic.

58.T F All procaryotic photosynthesizers are anoxygenic.

59.T F Fermentation involves the use of pyruvate and/or pyruvate derivatives as electron acceptors.

Answers to Sample Test Questions from Video 3

1.	True	7.	chemolithotrophy
2.	environment	8.	c
3.	DNA	9.	ensymes
4.	metabolism	10.	primary producers
5.	ATP	11.	sulfur; chemolithotrophs
6.	Photosynthesis	12.	a

Answers to Chapter 9 Sample Test Questions on Metabolism: The Generation of Energy

The question in this test bank have been rated for difficulty. The ratings run from 1 (for the easiest) to 8 (for the most difficult).

#	Answer / Diff.	#	Answer / Diff.
1	C / 3	29	C / 5
2	C / 5	30	coenzyme Q; cytochrome *c* / 4
3	C / 5	31	B / 5
4	D / 8	32	homolactic / 5
5	B / 5	33	heterolactic / 5
6	A / 5	34	D / 8
7	D / 5	35	hydrolysis; phosphorolysis / 7
8	C / 3	36	B / 1
9	C / 5	37	C / 5
10	C / 6	38	proteases / 4
11	F / 6	39	C / 6
12	B / 5	40	oxygen / 2
13	D / 8	41	B / 7
14	C / 3	42	chemiosmotic / 5
15	B / 3	43	conformational change / 5
16	A / 3	44	D / 6
17	Amphibolic / 7	45	T / 7
18	B / 5	46	T / 5
19	chemiosmotic;conformational change 6	47	T / 8
20	Pasteur effect / 6	48	T / 6
21	aerobic; anaerobic / 2	49	T / 5
22	B / 5	50	T / 3
23	D / 5	51	T / 3
24	B / 5	52	T / 7
25	C / 8	53	T / 7
26	A / 5	54	F / 7
27	A / 5	55	C / 8
28	A / 5	56	F / 5
		57	T / 7
		58	F / 7
		59.	T / 5

Answers to Sample Test Questions from Text Chapter 9 (Metabolism: The Generation of Energy) Web Site

1. metabolism
2. catabolism
3. Anabolism
4. All of the above
5. Is found in all major groups of microorganisms
6. True
7. Anaerobic fermentation
8. True
9. False
10. True

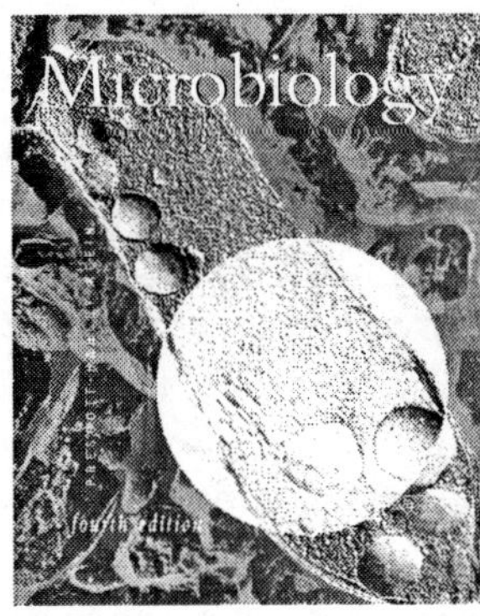

Microbial Genetics
Program 4
Reading the Code of Life

But the most important qualification of bacteria for genetic
studies is their extremely rapid rate of growth....a single
E. coli cell will grow overnight into a visible colony containing
millions of cells, even under relatively poor growth conditions.
Thus, genetic experiments on *E. coli* usually last one day, whereas
Experiments on corn, for example, take months. It is no
wonder we know so much more about the genetics of *E. coli* than about
the genetics of corn, even though we have been studying
corn much longer.

–R. F. Weaver and
P. W. Hendrick

One of microbiology's fifty most significant events in the past 125 years.

1953

James Watson and **Francis Crick**, together with **Maurice Wilkins** described the double-helix structure of
DNA. The chemical structure was based on X-ray crystallography of DNA done by Rosalind Franklin.
Crick, Wilkins, and Watson were awarded the Nobel Prize in Physiology or Medicine in 1962.

Student Assignments

- **Watch**   Video 4: Reading the Code of Life
- **Read**   Chapter 13: Microbial Genetics: General Principles in *Microbiology*, 4th ed,
  Prescott/Harley/Klein, McGraw-Hill
- **Lab Experiments** from *Laboratory Exercises in Microbiology*, 4th ed., Harley/Prescott,
  McGraw-Hill
  - Exercise 77: Bacterial Mutations
  - Exercise 78: Bacterial Transformations
  - Exercise 79: Bacterial Conjugation: The Transfer of Antibiotic Resistant
    Plasmids

Lesson Overview

Video 4: Reading the Code of Life
DNA is central to cell activity, carrying out the information for all proteins and replicating with great fidelity–except in the important case of mutations. Organisms also regulate the products made from genes in an effort to conserve energy and adapt to new environments.

Chapter 13: Microbial Genetics: General Principles
This chapter presents the basic concepts of molecular genetics: storage and organization of genetic information in the DNA molecule, mutagenesis, and repair. The role of microorganisms in screening procedures for mutagenic agents is also described. Primary emphasis is given to the genetics of bacteria.

Learning Objectives

After successfully completing this unit, the student will be able to:

- discuss the nature of the genetic code
- define a gene and discuss controlling elements, such as promoters and operators
- discuss the four parts (promoter, leader, coding region, trailer) associated with a bacterial gene
- discuss the nature and causes of mutations
- discuss the various genetic repair mechanisms and their limitations

Course Integrator

DNA as Genetic Material (p. 256)

Visual Resource Library	Discovery of Transformation Figure 13.1
Microbes in Motion II	Microbial Metabolism & Growth/Microbial Genetics
Video *Unseen Life on Earth*	Unit #4 Microbial Genetics: Mutations
Video *Unseen Life on Earth*	Unit #6 Microbial Evolution
Web Link	*Microbiology* Online Learning Center: http://www.mhhe.com/biosci/cellmicro/prescott/
Web Link	Converging on DNA 1900-1953 (http://www.gene.com:80/ae/AB/BC/1900-1953.html)

The Genetic Code (p. 258)

Gene Structure (p. 258)

Visual Resource Library	A Bacterial Structural Gene Figure 13.6
Transparency	A Bacterial Structural Gene Figure 13.6

Mutations and Their Chemical Basis (p. 262)

Microbes in Motion II	Control: Physical & Chemical/Physical Control/Radiation

Microbes in Motion II   Microbial Metabolism & Growth/Microbial Genetics
/Microbial Mutation

Detection and Isolation of Mutants (p. 269)
Visual Resource Library   Replica Plating Figure 13.15
Transparency   Replica Plating Figure 13.15

DNA Repair (p. 272)
Visual Resource Library   Recombination Repair Figure 13.19
Visual Resource Library   SOS Repair Figure 13.20

Microbial terms Used in Video 4 in Order of Appearance with Page Indexing to the Textbook (*Microbiology*, 4th ed., Prescott/Harley/Klein, McGraw-Hill)

DNA 423
base 213
nucleotides 203
transcribed 221
transcription 220
mRNA 242
ribosomes 228

protein synthesis 226
mutations 262
antibiotic resistance
   270
tuberculosis 773
colonize 584
penicillin 680

cell wall   208
radiation 130
plasmids 281
Euglena   546
represson 243
RNA polymerase 222
inducer 243
feedback inhibition 241

Web Links Related to Video 4 (Reading the Code of Life) and Readings in Chapter 13 (Microbial Genetics: General Principles) (Go to www.mhhe.com/prescott, Student on Line learning Center, Double Click on Chapter 13, Double Click on Web Sites)

Mutations of Chromosomes During Replication from the Graphics Gallery at Access Excellence.
http://www.acessexcellence.org/ab/gg/mutation.html

Experimental Demonstrations that DNA is the Genetic Material from the Graphics Gallery at
Access Excellence
http://www.accessexcellence.org/ab/gg/genematerial.html

1900-1953–Converging on DNA (Access Excellence)
http://www.accessexcellence.org/ab/bc/1900-1953.html

Microbial Terms Used in Chapter 13 (Microbial Genetics: General Principles)

Ames test 271
apurinic site 265
apyrimidine site 265
auxotrophs 263
back mutation 265
base analog 265
cistron 260
code degeneracy 258
coding region 262
codon 258
conditional mutation 263
frameshift 264
frameshift mutation 267
gene 258
genome 256

genotype 256
hypermutation 263
intercalating agent 265
leader sequence 261
missence mutation 267
mutagen 263
mutation 262
nonsense mutation 267
phenotype 256
photoreactivation 272
point mutation 267
promoter 260
proofreading 272
prototrophs 263
reading frame 260
recA protein 273
recombinant repair 273

replica plating 269
reverse mutation 265
sense codons 258
sense strand 260
Shine-Dalgarno sequence 262
silent mutation 267
SOS repair 273
specific mispairing 265
stop or nonsense codons 258
strain or clone 256
suppressor mutation 266
terminator sequence 262
transformation 256
transition mutation 264
transverse mutation 264
wild type 265
wobble 258

Sample Test Questions from Video 4 (Reading the Code of Life)

1. How many different nucleotides are their in DNA?
   a. One
   b. Two
   c. Three
   d. Four
   e. Five

2. When a message is synthesized from DNA to mRNA this is termed ___________.

3. DNA translation occurs on/in the
   a.. Ribosomes
   b. Nucleus
   c. Mitochondrion
   d. Lysosomes
   e. Golgi complex

4. An abnormal change that occurs in a DNA sequence is termed a ___________.

5. The more genetic variety, the better a species chances are of survival on the planet Earth.
   a. True
   b. False

6. Antibiotic resistance is a good example of genetic mutations.
   c. True
   d. False

7. The antibiotic penicillin kills bacteria by
   a. Destroying their nucleus
   b. Destroying their plasma membranes
   c. Destroying their cell walls
   d. All of the above

8. Most people really need antibiotics.
   a. True
   b. False

9. Most of the time cells have enzymes that can detect errors in DNA sequences and repair them.
   a. True
   b. False

10. *Euglena* is a
   a. Bacteria
   b. Virus
   c. Protists
   d. Fungus

Sample Test Questions from Text Chapter 13 (Microbial Genetics: General Principles) Web Site

1. Most mutations have coding regions and trailers.
   O True
   O False

2. Mutations are stable, heritable alterations in the gene sequence that
   O always produce phenotypic changes.
   O usually produce phenotypic changes.
   O rarely produce phenotypic changes.

3. To produce mutations, nucleic acids are altered
   O spontaneously
   O by radiation
   O by chemical mutagens
   O by any of the above

4. It is extremely important for an organism to keep the nucleotide sequence constant.
   O True
   O False

5. Often more than one repair system can correct a particular type of mutation.
   O True
   O False

6. Repair mechanisms always repair nucleotide sequences completely.
   O True
   O False

7. The collection of observable characteristics in an organism is called it's
   O genome
   O genotype
   O phenotype

8. An organism's specific set of genes is called it's
   O genome
   O genotype
   O phenotype

9. The knowledge that DNA is the genetic material for cells came from studies on virulent and nonvirulent bacterial pathogens.
   O True
   O False

10. A gene may be defined as the nucleic acid sequence that codes for
    O a polypeptide
    O tRNA
    O rRNA
    O all of the above

Chapter 13 Sample Test Questions on Microbial Genetics: General Principals

1. The principle of transformation was first demonstrated by which of the following individuals or groups of individuals?
   a.   Avery, MacLeod, and McCarty
   b   Griffith
   c.   Hershey and Chase

2. Which of the following individuals or groups of individuals demonstrated that the transforming principle was DNA?
    a. Avery, MacLeod, and McCarty
    b. Griffith
    c. Hershey and Chase
    d. Watson and Crick

3. The genetic code is said to be ___________ because more than one codon will specify a particular amino acid.
    a. ambiguous
    b. degenerate
    c. multiplicative
    d. repetitious

4. Which of the following is not considered the final product of the expression of a gene?
    a. a polypeptide chain
    b. an mRNA molecule
    c. a tRNA molecule
    d. an rRNA molecule

5. Key sequences exist within promoters, which vary somewhat among various promoters but are sufficiently constant that they can be represented by a sequence of bases most often found at each position. These representations are called ___________ sequences.
    a. convergence
    b. idealized
    c. consensus
    d. common

6. As the result of exposure to a mutagen, cytosine is substituted for thymine in one strand of DNA. Upon subsequent DNA replication, one of the daughter cells will have a GC pair in this position instead of an AT pair. This is called a(n) ___________ mutation.
    a. transversion
    b. transition
    c. frameshift
    d. insertion

7. Which of the following can lead to transition mutations?
    a. incorporation of a base analog that exhibits different base-pairing properties from thee base it replace
    b. chemical modification of an existing base in the DNA so that in the next round of replication it will pair differently from the unmodified base
    c. Either of these answers is correct.
    d. Neither of these answers is correct.

8. Which of the following is considered a reverse mutation that will restore the wild type phenotype?
   a. true reversion back to the wild type base sequence
   b. mutation to a different base sequence, but one that restores the amino acid sequence in the protein to the wild type sequence
   c. a mutation that restores the function of a protein even though it   does not restore the base sequence or the amino acid sequence to the wild type
   d. All of these.

9. Scientists study mutations in order to
   a. understand more about the nature of genes.
   b. understand more about the biochemistry of a particular microorganism.
   c. test the carcinogenic potential of a compound by determining the mutagenic capability of it.
   d. All of these.

10. The coding sequence in the DNA of __________ is normally continuous; that is, it is not interrupted by noncoding sequences.
    a. procaryotes
    b. eucaryotes
    c. both procaryotes and eucaryotes
    d. neither procaryotes nor eucaryotes

11. Which of the following types of mutation may play an important role in driving evolution because they are often nonlethal and, therefore, remain in the gene pool?
    a. nonsense mutations
    b. missense mutations
    c. both nonsense and missense mutations
    d. neither nonsense nor missense mutations

12. Which one of the following does not contribute to the utility of the strains of Salmonella typhimurium used in the Ames reversion assay?
    a. They are highly permeable to test substances.
    b. They are defective in excision repair and, therefore, cannot readily repair damage done by the test substances.
    c. They have plasmid genes that enhance error-prone repair, thereby increasing mutagenic sensitivity even though some damage is being repaired.
    d. All of these contribute to the utility of these strains.

13. Repair of thymine dimers using light to split the dimers apart into separate monomers is called
    a. photodedimerication.
    b. photoreactivation.
    c. photoreparation.
    d. None of these.

14. Repair of thymine dimers by splitting them apart is
    a. very error-prone.
    b. mildly error-prone.
    c. relatively error free.
    d. unpredictable with regard to error.

15. Mutations that inactivate metabolic pathways are called __________ mutations.
    a. morphological
    b. biochemical
    c. lethal
    d. conditional

16. Microbial strains that can grow on minimal medium are called __________.
    a. autotrophs
    b. auxotrophs
    c. prototrophs
    d. minitrophs

17. Mutant strains that are unable to grow on a minimal medium but require that the medium contains an adequate supply of the end product of a metabolic pathway are called :
    a. autotrophs
    b. auxotrophs
    c. prototrophs
    d. minitrophs

18. Mutations that result in the death of an organism when expressed are called __________ mutations.
    a. dominant
    b. recessive
    c. lethal
    d. conditional

19. Lethal mutations can be recovered in diploid organisms if they are
    a. dominant.
    b. recessive.
    c. conditional.
    d. None of these.

20. Lethal mutations can be recovered in haploid organisms if they are
    a. dominant.
    b. recessive.
    c. conditional.
    d. None of these.

21. Which of the following is in the coding region of a gene?
    a. promoter
    b. leader
    c. trailer
    d. None of these.

22. The strand of DNA containing the coding information is the __________ strand.
    a. sense
    b. template
    c. either sense or template
    d. neither sense nor template

23. Which of the following is not a termination codon?
    a. AUG
    b. UAA
    c. UAG
    d. UGA

24. Which of the following is the initiator codon?
    a. AUG
    b. UAA
    c. UAG
    d. UGA

25. In order to accommodate all 20 amino acids, it is necessary to read the base sequence in groups of __________.
    a. two
    b. three
    c. four
    d. five

26. A __________ mutation is one that causes premature termination of the synthesis of the protein product.
    a. missense
    b. nonsense
    c. silent
    d. None of these.

27. A mutation that changes the amino acid sequence of the resulting protein by substitution is called a __________ mutation.
    a. missense
    b. nonsense
    c. silent
    d. None of these.

28. A ___________ mutation does not alter the amino acid sequence of the resulting protein.
    a. missense
    b. nonsense
    c. silent
    d. None of these.

29. The transcribed and translated region of a gene is the ___________ region.
    a. leader
    b. promoter
    c. coding
    d. trailer

30. The region at which the RNA polymerase binds is the ___________ region.
    a. leader
    b. promoter
    c. coding
    d. trailer

31. The transcribed, but not translated, sequence that is immediately upstream of the region and encodes the functional product is called the ___________ region.
    a. leader
    b. promoter
    c. coding
    d. trailer

32. Which of the following is the most error-prone of the repair mechanisms?
    a. postreplication repair
    b. recombination repair
    c. SOS repair
    d. All are equally error-prone.

33. A mutation from the most prevalent form of a gene to a mutant form is called a ___________ mutation.
    a. forward
    b. reverse
    c. wild card
    d. escalating

34. A mutagen that inserts between the stacked bases of a DNA double helix, distorting the DNA to induce single-pair insertions or deletions is called a(n) ___________ agent.
    a. insertional
    b. interspersing
    c. intercalating
    d. distortional

35. Spontaneous mutations are caused by
    a. errors in DNA replication.
    b. insertion of transposons.
    c. either errors in DNA replication or insertion of transposons.
    d. neither errors in DNA replication nor insertion of transposons.

36. Within the promoter, there exists a(n) ___________ site that serves as the site of initial association of the RNA polymerase.
    a. recognition
    b. binding
    c. initializing
    d. insertion

37. The genetic code is not ___________ because each codon specifies one and only one amino acid.
    a. ambiguous
    b. degenerate
    c. multiplicative
    d. repetitious

38. Mutations that produce measurable differences only when certain environmental criteria are met are called ___________ mutations.
    a. morphological
    b. environmental
    c. conditional
    d. incomplete

39. Mutations that are insertions or deletions of one or two base pairs are best described as ___________ mutations.
    a. missense
    b. frameshift
    c. nonsense
    d. escalating

40. T F The terminator codons are translated codons; that is, they specify amino acids to be inserted into the last position of a growing polypeptide chain.

41. T F The initiator codon is a translated codon; that is, it specifies an amino acid to be inserted in the first position of the growing polypeptide chain.

42. T F For most genes, coding information is found in both strands.

43. T F All genes on the same DNA molecule use the same strand of DNA for their coding information.

44. T F The genetic code is usually translated in a non overlapping fashion.

45.T F The coding sequences in interrupted genes are called exons.

46.T F The noncoding sequences in interrupted genes are called introns.

47. A recipient cell that is temporarily diploid for a portion of the genome during the replacement process is called a(n) __________.

48. A second frameshift mutation that is downstream from a first mutation and restores the reading frame, thereby minimizing the phenotypic effect, is called a(n) __________ suppressor mutation.

49. Hershey and Chase demonstrated that when the bacteriophage T2 infected its host cell, the __________ was injected into the host but the __________ remained outside.

50. The __________ is the most prevalent form of a gene in a population.

51. __________ are physical or chemical agents that cause mutation.

52. The term __________ refers to the fact that the sequence of nucleotides in the DNA has a linear relationship to the sequence of amino acid in the protein encoded by that DNA sequence.

53. __________ mutations occur in the promoter or operator region of a gene or set of genes and affects the expression of the downstream genes without affecting the amino acid sequences of the gene products.

54. __________ mutations affect only a single base pair in a gene.

55. Mutations caused by some physical or chemical agent are called __________ mutations.

56. The __________ is the specific set of genes an organism possesses.

57. The __________ is the collection of characteristics of an organism that   an investigator observes.

58. A population of cells that are genetically identical is called a __________.

59. A consensus sequence in the untranslated leader sequence of a procaryotic mRNA which serves as a ribosome binding site is called the __________ sequence.

60.T F Eucaryotic genes are usually interrupted. Interesting exceptions   are the genes for histones which contain no introns.

61. The term __________ refers to the abnormally high rate of mutation that results from the activation of specific mutator genes.

Answers to Sample Test Questions from Video 4

1.	D	6.	True
2.	transcription	7.	C
3.	A	8.	False
4.	Mutation	9.	True
5.	True	10.	C

Answers to Sample Test Questions from Text Chapter 13 (Microbial Genetics: General Principles) Web Site

1.	True	6.	True
2.	Usually produce phenotypic changes	7.	Phenotype
3.	By any of the above	8.	Genotype
4.	True	9.	True
5.	True	10.	All of the above

Answers to Chapter 13 Sample Test Questions on Microbial Genetics: General Principles

The questions in this test bank have been rated for difficulty. The ratings run from 1 (for the easiest) to 8 (for the most difficult).

#	Answer/Diff.	#	Answer/Diff.
1	B / 5	19	B / 5
2	A / 5	20	C / 5
3	B / 8	21	D / 5
4	B / 8	22	C / 5
5	C / 6	23	A / 4
6	B / 6	24	A / 3
7	C / 6	25	B / 3
8	D / 6	26	B / 5
9	D / 4	27	A / 5
10	A / 5	28	C / 5
11	B / 7	29	C / 2
12	D / 8	30	B / 5
13	B / 5		
14	C / 5		
15	B / 4		
16	C / 4		
17	B / 4		
18	C / 1		

31	A/5
32	C/8
33	A/4
34	C/8
35	C/6
36	A/5
37	A/6
38	C/6
39	B/6
40	F/4
41	T/4
42	F/3
43	F/3
44	T/4
45	T/4
46	T/4
47	merozygote/5
48	intragenic/6
49	DNA; protein/5
50	wild type/6
51	Mutagens/5
52	colonearity/4
53	Regulatory/6
54	point/5
55	Induced/4
56	Genotype/4
57	Phenotype/4
58	Clone/5
59	Shine-Dalgaro/8
60	T/6

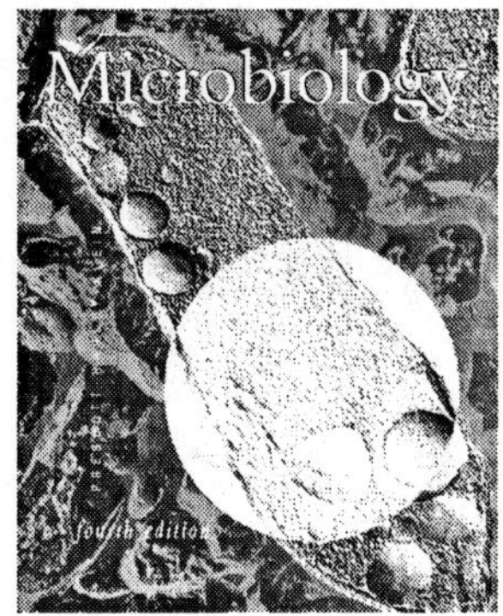

Microbial Genetics
Program 5
Genetic Transfer

The recombinant DNA breakthrough has provided us with a
new and powerful approach to the questions that have
intrigued and plagued man for centuries.

–Paul Berg

One of microbiology's fifty most significant events in the past 125 years.

1952

In this year, **Joshua Lederberg** and **Norton Zinder** reported on transduction, or transfer of genetic information by viruses. They showed for the first time that a phage of *Salmonella typhimurium* can carry DNA from one bacterium to another.

Student Assignments

- **Watch**   Video 5: Genetic Transfer
- **Read**   Chapter 14: Microbial Genetics: Recombination and Plasmids and
  Chapter 15: Recombinant DNA Technology in *Microbiology*, 4th ed.,
  Prescott/Harley/Klein, McGraw-Hill.
- **Lab Experiments** from *Laboratory Exercises in Microbiology*, 4th ed., Harley/Prescott,
  McGraw-Hill.
  Exercise 78: Bacterial Transformations
  Exercise 79: Bacterial Conjugation: The Transfer of Antibiotic-resistant Plasmids

Lesson Overview

Video 5: *Genetic Transfer*
Microbial populations achieve genetic diversity through horizontal gene transfer. Bacteria may transfer genes from one to another by conjugation, transformation, or transduction. Scientists often exploit these processes through recombinant DNA.

Chapter 14: Microbial genetics: Recombination and plasmids
This chapter begins with a general discussion of bacterial recombination, plasmids, and transposable elements, and then examines the acquisition of genetic information by conjugation, transformation, and transduction. The way these recombination procedures are used to map the bacterial genome is explained. Finally, viral recombination and genome mapping are discussed.

Chapter 15: Recombinant DNA Technology
This chapter focuses on practical applications of microbial genetic principles discussed in previous chapters. Although we have been altering the genetic makeup of organisms for centuries and nature has been doing it even longer, only recently have we been able to manipulate the DNA directly using genetic engineering or recombinant DNA technology. The potential benefits of these techniques are great and affect such diverse areas as medicine, agriculture, and industry. However, the use of these techniques is not without risks, and these risks must be considered in any discussion of this technology.

Learning Objectives

After successfully completing this unit, the student will be able to:

- discuss the nature of procaryotic recombination
- discuss the three ways (conjugation, transformation, and transduction) that bacteria acquire new genetic material
- discuss how plasmids and transposable elements can move genetic material between bacterial chromosomes and within a chromosome to cause changes in the genome and the phenotype of the organism
- discuss the use of these gene transfer procedures to map the bacterial genome
- discuss the sequencing of microbial genomes
- discuss the recombination that occurs when two viruses simultaneously infect the same host
- discuss the use of recombinant DNA technology to genetically engineer various organisms
- discuss the key role played by restriction endonucleases in genetic engineering
- discuss how plasmids, phages, and cosmids are used as vectors for insertion and expression of foreign genes in an organism
- discuss the use of both procaryotes and eucaryotes as target organisms for foreign gene insertion
- discuss the contributions already made by the use of this technology
- discuss the risks and the ethical problems associated with the use of this technology

Course Integrator

Chapter 14
Bacterial Recombination: General Principles (p. 279)

Visual Resource Library	Nonreciprocal recombination Figure 14.3
Visual Resource Library	Bacterial Conjugation Figure 14.4
Transparency	Merozygote Production and Fate Figure 14.4

Microbes in Motion II	Microbial Metabolism & Growth/
	Microbial Genetics/Recombination
Web Link	*Microbiology* Online Learning Center:
	http://www.mhhe.com/biosci/cellmicro/prescott/

Bacterial Plasmids (p. 281)

Visual Resource Library	F Plasmid Integration Figure 14.7
Transparency	F Plasmid Integration Figure 14.7
Microbes in Motion II	Microbial Genetics/Recombination/
Microbes in Motion II	Antimicrobial Resistance
	/Transfer & Spread of Resistance Genes

Transposable Elements (p. 285)

Visual Resource Library	R plasmids and transposons Figure 14.11
Microbes in Motion II	Antimicrobial Resistance/
	Transfer and Spread of Resistance Genes/
	/Transposons

Bacterial Conjugation (p. 289)

Visual Resource Library	Evidence for Conjugation Figure 14.12
Visual Resource Library	Evidence for Conjugation Figure 14.13
Visual Resource Library	Bacterial Conjugation Figure 14.14
Transparency	Bacterial Conjugation Figure 14.14
Transparency	F' Conjugation Figure14.15
Visual Resource Library	F' Conjugation Figure 14.15
Microbes in Motion II	Antimicrobial Resistance/
	Transfer and Spread of Resistance Genes/

DNA Transformation (p. 290)

Visual Resource Library	Bacterial Transformation Figure 14.16
Transparency	Bacterial Transformation Figure 14.16
Visual Resource Library	Mechanism of Transformation Figure 14.17
Transparency	Mechanism of transformation Figure 14.17

Transduction (p. 294)

Visual Resource Library	Lambda Phage Life Cycles Figure 14.18
Transparency	Lambda Phage Life Cycles Figure 14.18
Transparency	Generalized Transduction Figure 14.19

Chapter 15
Historical Perspectives (p. 311)

Visual Resource Library	Southern Blotting Technique Figure Number 15.5
Transparency	Southern Blotting Technique Figure 15.5
Microbes in Motion II	Microbial Metabolism & Growth/
	Microbial Genetics/Genetic Engineering
Web Link	*Microbiology* Online Learning Center
	http://www.mhhe.com/biosci/cellmicro/prescott/

Video *Unseen Life on Earth* — Unit #5 Genetic Engineering
Web Link — Access Excellence- About BioTech
(http://www.gene.com/ae/AB/)

Synthetic DNA (p. 314)
Visual Resource Library — Site-Directed Mutagenisis Figure 15.7

The Polymerase Chain Reaction (p. 315)
Visual Resource Library — The Polymerase Chain Reaction Figure 15.8
Transparency — The Polymerase Chain Reaction Figure 15.8
Microbes in Motion II — Virology/Viral Detection/Direct Detection Nucleic Acid
Hyperclinic — Direct Examination: Nucleic Acid Detection

Preparation of Recombinant DNA (p. 318)
Visual Resource Library — cDNA Synthesis Figure 15.9
Visual Resource Library — Recombinant Plasmid Production Figure 15.11
Transparency — Recombinant Plasmid Production Figure 15.11

Cloning Vectors (p. 324)
Visual Resource Library — Cloning with Plasmid Vector Figure 15.14
Transparency — Cloning with Plasmid Vectors Figure 15.14
Visual Resource Library — Lambda Phage Use as a Vector Figure 15.15
Transparency — Lambda Phage Use as a Vector Figure 15.15
Visual Resource Library — Recombinant Plasmid Detection Figure 15.17

Inserting Genes into Eucaryotic Cells (p. 326)

Expression of Foreign Genes in Bacteria (p. 326)

Applications of Genetic Engineering (p. 328)
Web Link — Biotech Applied (Access Excellence):
www.gene.com/ae/AB/BA/
Web Link — Use of DNA in Identification
(http://esg-www.mit.edu:8001/esgbio/rdna/landerfinger.html)
Web Link — DNA Fingerprinting in Human Health & Society

Historical Perspectives (p. 311)
Visual Resource Library — Southern Blotting Technique Figure Number 15.5
Transparency — Southern Blotting Technique Figure 15.5
Microbes in Motion II — Microbial Metabolism & Growth/
Microbial Genetics/Genetic Engineering
Web Link — *Microbiology* Online Learning Center
http://www.mhhe.com/biosci/cellmicro/prescott/
Video *Unseen Life on Earth* — Unit #5 Genetic Engineering
Web Link — Access Excellence- About BioTech
(http://www.gene.com/ae/AB/)

Synthetic DNA (p. 314)

Visual Resource Library	Site-Directed Mutagenisis Figure 15.7

The Polymerase Chain Reaction (p. 315)

Visual Resource Library	The Polymerase Chain Reaction Figure 15.8
Transparency	The Polymerase Chain Reaction Figure 15.8
Microbes in Motion II	Virology/Viral Detection/Direct Detection Nucleic Acid
Hyperclinic	Direct Examination: Nucleic Acid Detection

Preparation of Recombinant DNA (p. 318)

Visual Resource Library	cDNA Synthesis Figure 15.9
Visual Resource Library	Recombinant Plasmid Production Figure 15.11
Transparency	Recombinant Plasmid Production Figure 15.11

Cloning Vectors (p. 324)

Visual Resource Library	Cloning with Plasmid Vector Figure 15.14
Transparency	Cloning with Plasmid Vectors Figure 15.14
Visual Resource Library	Lambda Phage Use as a Vector Figure 15.15
Transparency	Lambda Phage Use as a Vector Figure 15.15
Visual Resource Library	Recombinant Plasmid Detection Figure 15.17

Inserting Genes into Eucaryotic Cells (p. 326)

Expression of Foreign Genes in Bacteria (p. 326)

Applications of Genetic Engineering (p. 328)

Web Link	Biotech Applied (Access Excellence): www.gene.com/ae/AB/BA/
Web Link	Use of DNA in Identification (http://esg-www.mit.edu:8001/esgbio/rdna/landerfinger.html)
Web Link	DNA Fingerprinting in Human Health & Society

Microbial Terms Used in Video 5 in Order of Appearance with Page Indexing to the Textbook (*Microbiology*, 4th ed., Prescott/Harley/Klein, McGraw-Hill.)

transformation 290	diphtheria 767	transposition 285
transduction 294	toxin 284	transposon 285
conjugation 289	bacteriophage 356	parasite 581
horizontal gene transmission 279	sex pilus 283	*Agrobacterium* 464
	plasmid 281	transgenic plants 330
virulence traits 731	lytic cycle 357	

Web Links related to Video 5 (Genetic Transfer) and Readings in Chapter 14 (Microbial Genetics: Recombination and Plasmids) and Chapter 15 (Recombinant DNA Technology) (Go to www.mhhe.com/prescott. Student on Line Learning Center, Double Click on Chapter 14 [and then 15], Double Click on Web Sites)

Chapter 14

A New Gene map of the Human Genome. The International RH Mapping Consortium. *Comparative Scale of Mapping.* Graphics gallery at Access Excellence.
http://www.nebi.nlm.gov/genemap/

Understanding Gene Testing. National Cancer Institute.
http://www.acessexcellence.org/ab/gg/comparative.html

Chapter 15

Biotech Applied. Biotech Applied reviews the roles biotechnology has played in the fields of medicine, fuels, farms, forensics, food and the environment.
http://www.acessexcellence.org/ab/ba/

Polymerase Chain Reaction (PCR)
http://www.accessexcellence.org/ab/gg/polymerase.html

Inserting a DNA Sample into a Plasmid
http://www.accessexcellence.org/ab/gginserting.html

Bioethics. Includes: *Biotechnology's Impact on Society, Challenges to Public Policy, NIH Published Revised Guidelines for Research Involving Recombinant DNA Molecules* and other very topical social issues in biotechnology.
http://www.accessexcellence.org/ab/ie/bioethics.html

Microbial Terms Used in Chapter 14 (Microbial Genetics: Recombination and Plasmids)

abortive transductants 296
annotation 303
bacteriocin 284
competent 292
composite transposons 285
conjugation 289

conjugative plasmid 281
crossing over 279
curing 281
denaturation mapping 307
endogenote 280
episome 281

exogenote 280
F factor 282
F' plasmid 290
generalized transducing particle 296
generalized transduction 295

Microbial Terms Used in Chapter 15 (Recombinant DNA Technology)

Sample Test Questions from Video 5 (Genetic Transfer)

1. Which of the following is not a mechanism for microbial genetic recombination?
   a. Conjugation
   b. Transformation
   c. Budding
   d. Transduction

2. Human beings transmit their genetic information to subsequent generations by way of
   a. vertical gene transmission
   a. Horizontal gene transmission
   b. Vertical gene transmission
   c. Diagonal gene transmission
   d. All of the above

3. Antibiotic resistance is a good example of
   a. Horizontal gene transmission
   b. Vertical gene transmission
   c. Diagonal gene transmission
   d. All of the above

4. The small DNA fragments that move from one microbe to another are called __________.

5. The form of gene transfer and recombination in bacteria that requires direct cell-to-cell contact is called __________.

6. The transfer of genes between bacteria by bacteriophages is called __________.

7. The type of gene transfer in bacteria in which a piece of free DNA is taken up by a bacterial cell and integrated into the recipient gene is called __________.

8. The majority of microbes from the environment can be grown in the laboratory.
   a. True
   b. False

9. A zone of inhibition around a growing bacterial colony indicates resistance to the antibiotic.
   a. True
   b. False

10. The starchy root that sustains 500 million people in Africa is called __________.

Sample Test Questions from Text Chapter 14 (Microbial Genetics: Recombination and Plasmids) Web Site

1. Recombination is the process in which a new recombinant chromosome is formed by combining genetic material from two organisms.
   O True
   O False

2. The transfer of genetic material between bacteria in direct physical contact is called
   - O conjugation
   - O transformation
   - O transduction

3. The transfer of a naked piece of DNA between bacteria is called
   - O conjugation
   - O transformation
   - O transduction

4. The transport of bacterial DNA to other bacteria via bacteriophages is called
   - O conjugation
   - O transformation
   - O transduction

5. Transposable elements are
   - O plasmids
   - O viruses
   - O pieces of DNA

6. Genetic material is moved between bacterial chromosomes and within chromosomes by
   - O plasmids
   - O transposable elements
   - O plasmids and transposable elements

7. Recombination of virus genomes occurs
   - O by transformation
   - O by transduction
   - O when two viruses with homologous chromosomes infect a host cell simultaneously

8. Plasmids are small, circular DNA molecules that can exist independently of host chromosomes.
   - O True
   - O False

9. Plasmids are stably inherited.
   - O True
   - O False

10. Plasmids have proved invaluable to scientists in constructing and transferring new genetic combinations and in cloning genes.
    - O True
    - O False

11. Transposable elements
    - o   cause mutations
    - o   block translation
    - o   carry antibiotic resistance genes
    - o   all of the above

Sample Test Questions from Text Chapter 15 (Recombinant DNA Technology) Web Site

1. DNA can be cloned in host cells.
    - o   True
    - o   False

2. The production of recombinant DNA molecules depends on the ability of __________ to cleave DNA at specific sites.
    - o   Restriction endonucleases
    - o   Reverse transcriptases

3. An example of a vector used in genetic engineering is
    - o   a plasmid
    - o   a virus
    - o   a bacteriophage
    - o   all of the above can be vectors

4. Vectors can replicate within a host cell while carrying foreign DNA.
    - o   True
    - o   False

5. Vectors possess phenotypic traits that allow them to be recognized.
    - o   True
    - o   False

6. The southern blot technique is used to
    - o   detect a specific DNA sequence
    - o   engineer Ti plasmids
    - o   detect viral antigens

7. The polymerase chain reaction allows small amounts of DNA to be increased in concentration thousands of times.
    - o   True
    - o   False

8. Agarose gel electrophoresis is used to
   - O increase DNA
   - O separate DNA fragments by size
   - O separate DNA fragments by sequence.

9. The deliberate modification of an organism's genetic information by directly changing its nucleic acid genome is called
   - O cloning
   - O genetic insertion
   - O genetic engineering

10. The collection of methods used in genetic engineering is called recombinant DNA technology.
    - O True
    - O False

Chapter 14 Sample Test Questions on Microbial Genetics: Recombination and Plasmids

1. The formation of a new chromosome, different from either parent, by the combination of genetic material from two organisms is called
   a. transformation.
   b. conjugation.
   c. recombination.
   d. transduction.

2. A reciprocal exchange in which a pair of homologous sequences break and rejoin in a crossover is called __________ recombination.
   a. general
   b. site-specific
   c. replicative
   d. None of these.

3. The nonhomologous insertion of DNA into the chromosome is called __________ recombination.
   a. general
   b. site-specific
   c. replicative
   d. None of these.

4. The incorporation of a single strand to form a stretch of heteroduplex DNA is called __________ recombination.
   a. general
   b. site-specific
   c. replicative
   d. None of these.

5. A donor DNA that enters the bacterium by one of several mechanisms is called a(n):
   a. merozygote
   b. exogenote
   c. endogenote
   d. None of these.

6. The genome of a recipient cell is called a(n) __________.
   a. merozygote
   b. exogenote
   c. endogenote
   d. None of these.

7. A recipient cell that is temporarily diploid for a portion of the genome during the replacement process is called a(n) __________.
   a. merozygote
   b. exogenote
   c. endogenote
   d. None of these.

8. The elimination of a plasmid either spontaneously or by treatments that inhibit plasmid replication without affecting host cell reproduction is called
   a. transformation.
   b. conjugation.
   c. curing.
   d. plasmolysis.

9. Plasmids that have genes for resistance to antibiotics are called __________ factors.
   a. col
   b. resistance
   c. virulence
   d. metabolic

10. __________ plasmids make a bacterium pathogenic by conferring resistance to host defenses or by encoding the production of a toxin.
    a. Col
    b. Resistance
    c. Virulence
    d. Metabolic

11. Which of the following is not true about transposable elements?
    a. They move about the chromosomes within an organism.
    b. They lack the infectious cycle of temperate phages.
    c. They reproduce independently like plasmids.
    d. All of these are true about transposable elements.

12. Which of the following is not an effect of transposable elements?
    a. insertional mutagenesis
    b. termination of translation and/or transcription by insertion of stop codons
    c. insertional activation
    d. All of these are effects of transposable elements.

13. In an F+ X F- conjugation, the donor is the __________ strain.
    a. F+
    b. F-
    c. both F+ and F- (It is a reciprocal exchange.)
    d. neither F+ nor F- (There is no exchange.)

14. After an F+ X F- mating, the donor becomes __________ and the recipient becomes
    __________.
    a. F+; F-
    b. F-; F+
    c. F-; F-
    d. F+; F+

15. Which of the following matings can take place?
    a. F+ X F+
    b. Hfr X Hfr
    c. Hfr X F+
    d. None of these.

16. When an F plasmid integrates into the host chromosome, the strain is referred to as
    a. F+
    b. Hfr
    c   F'
    d. F-

17. Which of the following best represents the order of gene transfer in an Hfr X F- mating?
    a. all plasmid genes followed by some or all of the chromosome
    b. part of the plasmid followed by the chromosome followed by the rest of the plasmid
    c. the chromosome followed by the plasmid
    d. part of the chromosome followed by the plasmid followed by the rest of the chromosome

18. Complete transfer in an Hfr X F- mating takes approximately __________ minutes.
    a. 20
    b. 50
    c. 100
    d. 200

19. An F' plasmid results when
    a. an F+ X F- mating is interrupted before completion.
    b. an Hfr X F- mating is interrupted before completion.
    c. an integrated F plasmid is incorrectly excised, bringing host genes with it.
    d. None of these.

20. Which of the following matings can take place?
    a. F' X F-
    b. F' X Hfr
    c. F' X F+
    d. None of these.

21. When an F' plasmid acts as the donor in a mating, which of the following do(es) not happen?
    a. The recipient becomes F'.
    b. The plasmid is transferred, including the chromosomal genes now on the plasmid.
    c. Chromosomal genes not on the plasmid are transferred.
    d. All of these happen in an F' X F- mating.

22. F' X F- conjugation is called
    a. transduction.
    b. sexduction.
    c. F-duction.
    d. transformation.

23. When a recipient cell acquires a piece of naked DNA from the environment, it is called
    a. conjugation.
    b. transformation.
    c. transduction.
    d. transfection.

24. When bacterial genes are transferred to another bacterium by a virus, it is called
    a. conjugation.
    b. transformation.
    c. transduction.
    d. transfection.

25. Which of the following do(es) not occur with generalized transduction?
    a. degradation of the host chromosome into randomly sized fragments
    b. packaging of any DNA fragment of the appropriate size
    c. transfer of any bacterial gene to the subsequent host
    d. All of these occur with generalized transduction.

26. Which of the following is not true of specialized transduction?
    a. It is carried out by temperate phage.
    b. It is restricted to those genes on either side of an integrated prophage.
    c. The phage genome remains normal and carries the full complement of phage genes.
    d. All of these are true of specialized transduction.

27. The frequency of cotransfer of two genes can be used to construct ___________ maps
    indicating the distance of two genes from each other.
    a. transformation
    b. generalized transduction
    c. either transformation or generalized transduction
    d. neither transformation nor generalized transduction

28. Which of the following cannot be used to cure a bacterial strain of its plasmids?
    a. UV and/or ionizing radiation
    b. acridine mutagens
    c. thymine starvation
    d. All of these can be used to cure plasmids.

29. Proteins produced by bacteria that destroy other bacteria are generally called
    a. colicins.
    b. bacteriocins.
    c. bacteriolysins.
    d. plasmolysins.

30. The most efficient mechanism of gene transfer between bacteria is
    a. transduction.
    b. transformation.
    c. Hfr X F- conjugation.
    d. F+ X F- conjugation.

31. Who is credited with demonstrating the existence of mating types in bacteria?
    a. Hayes
    b. Lederberg
    c. Tatum
    d. None of these.

32. Transfer of genetic information via direct cell-cell contact is called
    a. transformation.
    b. transduction.
    c. transfection.
    d. conjugation.

33. Specialized transduction can be carried out by
    a. any bacteriophage.
    b. any temperate bacteriophage.
    c. only those temperate bacteriophages that integrate into the host chromosome.
    d. All of these.

34. When DNA from a wild type and a mutant strain are reannealed, homologous regions pair normally but mutant regions bubble; this process, known as ___________ mapping, can be used to map the location of the mutant genes.
    a. denaturation
    b. heteroduplex
    c. restriction endonuclease
    d. hybridoma

35. Which of the following is true of the integration of a viral genome into the host chromosome?
    a. It is a form of site-specific recombination.
    b. The enzymes are specific for the virus and its host.
    c. Both of these answers are true.
    d. Neither of these answers is true.

36. Which of the following is not a possible fate for an exogenote?
    a. integration into the host chromosome
    b. independent replication and functioning
    c. survival without replication
    d. All are possible fates of exogenotes.

37. Transposable elements that carry genes for functions other than those needed for transposition (e.g., antibiotic resistance) are called
    a. insertion sequences.
    b. composite transposons.
    c. either insertion sequences or composite transposons.
    d. neither insertion sequences nor composite transposons.

38. T F In an HFR × F+ mating the conjugation bridge usually breaks before chromosomal transfer is complete. Therefore, the recipient remains F-.

39. T F In the mechanics of conjugation, exclusive of gene transfer, Hfr and F+ strains behave the same.

40. T F When a genome is mapped with Hfr conjugation, a low frequency of gene transfer means that the gene is located close to the integration site of the plasmid.

41. T F The entire genome can be mapped because the conjugation bridge usually stays intact long enough for the entire chromosome to be transferred.

42. T F Transposable elements have been found only in procaryotes and do not appear to play a major role in eucaryotic genetics.

43. T F Each antibiotic resistance gene is carried on a single plasmid.

44. T F In an F+ X F- mating, there is usually a transfer of host chromosomal genes to the recipient.

45. T F Hfr strains can be used to map the entire bacterial genome by constructing overlapping maps from Hfr strains with different integration sites.

46. T F The conjugation bridge in an Hfr X F- mating usually breaks before chromosome transfer is complete; however, because at least part of the plasmid is transferred first, the recipient becomes F+.

47. T F When a genome is mapped with Hfr conjugation, a high frequency of gene transfer means that the gene is located close to the integration site of the plasmid.

48. T F Nonconjugative plasmids can be transmitted to another cell if conjugation is initiated by another (conjugative) plasmid.

49. T F Transformation has been observed in only a limited number of species in nature, but we have found ways to force the process upon other species in the laboratory.

50. A __________ is a latent form of a virus genome that remains within the host without destroying it.

51. A __________ is a piece of extrachromosomal DNA that has its own replication origin.

52. A plasmid that can either exist independently of the chromosome or be integrated into it is called a(n) __________.

53. Plasmids that have genes for pili and can transfer copies of themselves to other bacteria during conjugation are called __________ plasmids.

54. __________ is the process by which some cells are incapable of receiving an exogenote.

55. __________ plasmids carry genes for enzymes that degrade environmental substances such as aromatic compounds or pesticides.

56. In order to take up a naked DNA molecule, a cell must be __________, which may occur only at certain stages in the life cycle of the organism.

57. T F All resistance plasmids are nonconjugative.

58. Transfer of genes from one mature independent organism to another is called:
    a. horizontal gene transfer
    b. lateral gene transfer
    c. vertical gene transfer
    d. orthogonal gene transfer

59. Transfer of genes from parent to offspring is called
    a. horizontal gene transfer
    b. lateral gene transfer
    c. vertical gene transfer
    d. orthogonal gene transfer

60. The process of _______________ involves identifying open reading frames (ORFs), determining potential amino acid sequences, and comparing these to those of known proteins.

61. A _____________ is a long sequence of nucleotides that has been determined by analysis of overlaps between shorter sequenced fragments.

Chapter 15 Sample Test Questions on Recombinant DNA Technology

1. Restriction endonucleases were first described by
   a. Arber.
   b. Smith.
   c. both Arber and Smith.
   d. neither Arber nor Smith.

2. The first restriction endonuclease to be isolated was EcoRI. It was isolated by
   a. Arber.
   b. Smith.
   c. Boyer.
   d. Baltimore.

3. The enzyme reverse transcriptase was discovered by
   a. Temin.
   b. Baltimore.
   c. both Temin and Baltimore.
   d. neither Temin nor Baltimore.

4. Who of the following produced the first recombinant plasmid and introduced it into a bacterial host where it was replicated?
   a. Baltimore and Temin
   b. Cohen and Boyer
   c. Jackson, Symons, and Berg
   d. Arber and Smith

5. Who of the following developed procedures for rapidly sequencing DNA?
   a. Maxam
   b. Gilbert
   c. Sanger
   d. All of these.

6. A gene probe that is generated by isolating mRNA and using the enzyme reverse transcriptase to produce a DNA copy is called a(n)
   a. mRNA probe.
   b. cDNA probe.
   c. synthesized probe.
   d. cloned probe.

7. A gene probe that is produced by analyzing the amino acid sequence of all or part of a protein and chemically constructing a DNA sequence corresponding to the deduced gene is called a(n) ___________ probe.
   a. mRNA
   b. cDNA
   c. synthesized
   d. cloned

8. Which of the following is not true of cloning vectors?
   a. They are small well-characterized DNA molecules.
   b. They contain at least two replication origins.
   c. They can be replicated within an appropriate host.
   d. All of these are true of cloning vectors.

9. Which of the following cannot be used as cloning vectors?
   a. plasmids
   b. cosmids
   c. bacteriophages
   d. All of these can be used as cloning vectors.

10. Plasmid cloning vectors are usually introduced into bacterial hosts by
    a. conjugation.
    b. transformation.
    c. transduction.
    d. plasmolysis.

11. Plasmid vectors are often useful because they contain ___________ genes that can be used to select for recombinants.
    a. metabolic activation
    b. antibiotic resistance
    d. insertion sequence
    c. promoter/operator

12. In order to be expressed in procaryotes, eucaryotic genes must be associated with a bacterial
    a. promoter.
    b. leader sequence.
    c. both promoter and leader sequence.
    d. neither promoter nor leader sequence.

13. In order to express eucaryotic genes in a bacterium, the ___________ must first be removed.
    a. introns
    b. exons
    c. both introns and exons
    d. neither introns nor exons

14. Which of the following is (are) currently being produced by recombinant DNA technology
    for use in humans?
    a. human growth hormone
    b. human interferon
    c. human insulin
    d. All of these.

15. Which of the following best describes the basis for separation of DNA fragments during
    agarose gel electrophoresis?
    a. The fragments with the highest charge/mass ratio will migrate farthest.
    b.. The fragments with the smallest charge/mass ratio will migrate   farthest.
    c. The largest fragments will migrate farthest.
    d. The smallest fragments will migrate farthest.

16. A DNA molecule used to carry a foreign gene into a host organism is called a
    a. plasmid.
    b. vector.
    c. probe.
    d. blot.

17. A ___________ is a DNA molecule used in hybridization reactions to detect the presence of
    a particular gene in separated DNA fragments.
    a. plasmid
    b. vector
    c. probe
    d. blot

18. When a DNA probe is constructed using the amino acid sequence of a protein and the known
    genetic code to provide the information for the nucleotide sequence of the probe, the resulting
    probe will most likely be
    a. exactly the same as the naturally occurring sequence for the gene for that protein.
    b. sufficiently similar to the naturally occurring sequence for the gene for that protein that
        hybridization to a DNA fragment containing the naturally occurring sequence will occur.
    c. so different from the naturally occurring sequence for the gene for that protein that
        hybridization to a fragment of DNA containing the naturally occurring sequence will not
        occur.
    d. any of the above with roughly equal frequency of occurrence.

19. Complementary DNA (cDNA) probes are produced by
    a. restriction endonucleases.
    b. DNA polymerase.
    c. DNA ligase.
    d. reverse transcriptase.

20. An enzyme that cuts within a DNA molecule is a(n)
    a. exonuclease.
    b. endonuclease.
    c. ligase.
    d. methylase.

21. Genetically engineered interferon will most likely be used for
    a. a vaccine to prevent viral infections.
    b. a treatment against viral infections.
    c. a treatment against genetic defects.
    d. a treatment against bacterial infections.

22. Cloning a gene involves all of the following except
    a. isolating the fragment of DNA containing the desired gene.
    b. insertion of the gene into an appropriate vector.
    c. expression of the vector and the gene in a cell-free environment.
    d. All of these are involved in cloning a gene.

23. Antibiotics are usually incorporated into the culture medium in order to
    a. select against organisms that have not incorporated the plasmid.
    b. select against organisms that have incorporated a plasmid not containing the desired
       gene.
    c. Both of these answers are correct.
    d. Neither of these answers is correct.

24. Restriction endonucleases in bacteria probably evolved in order to
    a. carry out natural genetic engineering.
    b. protect the bacteria from infection by viruses.
    c. utilize nucleic acids as a food (energy) source.
    d. All of these.

25. When a eucaryotic gene is cloned into a bacterium, the advantage of a complementary DNA
    (cDNA) gene being used instead of fragments of genomic DNA is that
    a. the promoter and terminator are found in the cDNA gene but not in the genomic
       fragment.
    b. the introns have been removed from the cDNA gene but not from the genomic
       fragment.
    c. the cDNA is made with the nucleotides found in the procaryote but not in the
       eucaryote.
    d. There is no advantage to using a cDNA gene rather than a genomic fragment.

26. Which of the following hormones has not been produced by recombinant DNA technology?
    a. somatostatin
    b. estrogen
    c. human growth hormone
    d. insulin

27. Recombinant DNA technology is currently being used (or is under investigation for future use for
    a. production of synthetic vaccines.
    b. diagnosis of certain genetic disorders.
    c. somatic gene therapy for certain genetic disorders.
    d. All of these.

28. A(n) __________ vector is a plasmid that can be replicated in several different organisms because it has at least one origin of replication that will function in each host.
    a. shuttle
    b. chimeric
    c. expression
    d. multi

29. A(n) __________ vector has all the necessary transcription and translation start and stop signals with useful restriction endonuclease cleavage sites nearby so that foreign DNA fragments can be inserted in proper orientation for immediate production of the gene product.
    a. shuttle
    b. chimeric
    c. expression
    d. multi

30. Restriction endonucleases
    a. make a blunt cut on the two DNA strands so that there are no single-strand regions.
    b. make staggered cuts on the DNA so that single-strand ends are formed that can be used to insert foreign DNA cut with the same enzyme.
    c. Some make a blunt cut on the two DNA strands so that there are no single-strand regions and some make staggered cuts on the DNA so that single-strand ends are formed that can be used to insert foreign DNA cut with the same enzyme.
    d. The same enzyme both makes a blunt cut on the two DNA strands so that there are no single-strand regions and makes staggered cuts on the DNA so that single-strand ends are formed that can be used to insert foreign DNA cut with the same enzyme depending upon the incubation conditions.

31. One of the most widely used plasmid vectors is
    a. lambda.
    b. T4 DNA ligase.
    c. pBR322.
    d. SV-40.

32. Restriction endonucleases are most useful for DNA cloning if the plasmid vector has
__________ cleavage site(s) for that particular enzyme.
    a. only one
    b. one or two
    c. many
    d. The number of cleavage sites is not important.

33. Recombinant DNA technology does not rely on which of the following enzymes?
    a. restriction endonucleases
    b. DNA methylase
    c. DNA ligase
    d. Recombinant DNA technology relies on all of these.

34. Movement of charged molecules in an electrical field, which is used to separate nucleic acid
fragments for recombinant DNA work, is called
    a. iontophoresis.
    b. nucleophoresis.
    c. electrophoresis.
    d. plasmaphoresis.

35. A(n)__________ __________ is a piece of DNA with all of the features necessary for
chromosomal replication and which can carry large (100kb-2000kb) pieces of foreign DNA
into a host organism.

36. Which of the following is not true about the use of nonradiolabelled hybridization probes when
compared to radiolabelled probes?
    a. nonradiolabelled probes are faster to be detected
    b. nonradiolabelled probes are safer to use
    c. nonradiolabelled probes are more sensitive than radiolabelled probes
    d. All of these are true.

37. Gene copies with known sequence alterations are normally produced by a process called
    a. polymerase chain reaction.
    b. site-directed mutagenesis.
    c. genomic library mutagenesis.
    d. None of these are correct.

38. The polymerase chain reaction (PCR) can be used to produce __________ of copies in 1 or 2
hours.
    a. hundreds
    b. thousands
    c. millions
    d. billions

39. An enzyme that cuts DNA from one end or another one nucleotide at a time is a(n)
    a. exonuclease
    b. endonuclease
    c. ligase
    d. methylase

40. Which of the following is a goal for the use of recombinant DNA technology in agriculture?
    a. introduce nitrogen-fixation capability into crop plants that currently do not have them
    b. protect crop plants against frost damage and other environmental stress
    c. make crop plants poisonous to insect pests
    d. All of these are goals for the use of recombinant DNA technology in agriculture.

41. Which of the following has been used to introduce genes into eucaryotic cells?
    a. microinjection
    b. electroporation
    c. firing microprojectiles into target cells with a gene gun
    d. All of these.

42. T F Cosmids are plasmids that can be packaged into capsids of the bacteriophage lambda; therefore, they can be transmitted like phages, but they can exist and replicate in a cell like plasmids.

43. T F Transposons are frequently used as cloning vectors.

44. T F The Southern blotting technique for transferring DNA from an agarose gel to a piece of nitrocellulose membrane is named for E. M. Southern, the person who developed the procedure.

45. T F Cosmids are so named because they can be used to express foreign genes in a variety of different hosts.

46. T F Genetic engineering methods have been used to produce vaccines.

47. T F Some DNAs have been modified in order to protect them from the action of a particular restriction endonuclease. This usually results in the DNA being resistant to most other restriction endonucleases as well.

48. T F Recombinant DNA technology has been used to develop a strain of bacteria that can be useful in cleaning up oil spills.

49. T F One of the major fears about the use of recombinant DNA technology is that ecosystem disruption could result from environmental release of genetically engineered organisms.

50. T F Some plasmid vectors have incorporated the regulatory sequences of the lactose operon so that the expression of the recombinant gene can be induced at the appropriate time.

51.T F Regardless of the exact approach taken to recombinant DNA technology, one of the keys to successful cloning is choosing the right vector.

52.T F One of the major advantages to using plasmids as cloning vectors is that the plasmid can be readily separated from the bacterial chromosome.

53.T F When a eucaryotic gene is cloned into a bacterium, the eucaryotic regulatory sequences should be maintained in order to achieve maximum expression of the gene.

54.T F It is necessary to remove introns from eucaryotic genes before cloning them into a procaryotic organism because they cannot be removed from the RNA transcript by the same kind of processing that normally occurs in the eucaryotic cell of origin.

55. A genomic ___________ is a set of cloned fragments that includes the entire genome of an organism.

56.T F Genetically engineered mice can now produce fully human monoclonal antibodies.

57. The first FDA approved genetically engineered food to be marketed is the ___________.

58. The forensic science of DNA fingerprinting includes:
    a. restriction endonuclease digestion of DNA samples and electrophonetic analysis of the fragments
    b. polmerase chain reaction (PCR) amplification of certain DNA sequences that differ between individuals
    c. both a and b are correct
    d. neither a nor b is correct

59. Which of the following types of cloning vector cancarry the least amount of foreign DNA?
    a. artificial chromosome
    b. bacteriophage
    c. cosmid
    d. plasmid

60. Which of the following types of cloning vector can carry the largest amount of foreign DNA?
    a. artificial chromosome
    b. bacteriophage
    c. cosmid
    d. plasmid

Answers to Sample Test Questions from Video 5

1. C
2. B
3. A
4. Plasmids
5. Conjugation

6. Transduction
7. Transformation
8. False
9. False
10. Cassava

Answers to Sample Test Questions from Text Chapter 14 (Microbial Genetics: Recombination and Plasmids) Web Site

1. True
2. Conjugation
3. Transduction
4. Transformation
5. Pieces of DNA

6. Plasmids and transposable elements
7. When two viruses with homologous chromosomes infect a host cell simultaneously
8. True
9. True
10. True
11. All of the above

Answers to Sample Test Questions from Text Chapter 15 (Recombinant DNA Technology) Web Site

1. True
2. Restriction endonucleases
3. All of the above can be vectors
4. True
5. True

6. Detect a specific DNA sequence
7. True
8. Separate DNA fragments by size
9. Genetic engineering
10. True

Answers to Chapter 14 Sample Test Questions on Microbial Genetics: Recombination and Plasmids

The questions in this test bank have been rated for difficulty. The ratings run from 1 (the easiest) to 8 (for the most difficult)

#	Answer/Diff.	#	Answer/Diff.
1	C / 5	5	B / 5
2	A / 5	6	C / 5
3	B / 5	7	A / 5
4	D / 6	8	C / 5

9	B / 3		56	competent / 7
10	C / 5		57	F / 5
11	C / 6		58	A / 4
12	D / 6		59	C / 4
13	A / 5		60	annotation / 7
14	D / 7		61	contig / 8
15	D / 3			
16	B / 4			
17	B / 8			
18	C / 6			
19	C / 8			
20	A / 5			
21	C / 7			
22	B / 7			
23	B / 4			
24	C / 4			
25	D / 5			
26	C / 6			
27	C / 5			
28	D / 6			
29	B / 5			
30	C / 8			
31	A / 6			
32	D / 4			
33	C / 6			
34	B / 5			
35	C / 6			
36	D / 7			
37	B / 5			
38	T / 4			
39	T / 5			
40	F / 4			
41	F / 6			
42	F / 6			
43	T / 3			
44	F / 5			
45	T / 6			
46	F / 5			
47	T / 4			
48	T / 5			
49	T / 6			
50	prophage / 6			
51	plasmid / 4			
52	episome / 6			
53	conjugative / 5			
54	Host restriction / 7			
55	Metabolic / 4			

Answers to Chapter 15 Test Questions on Recombinant DNA Technology

The questions in this test bank have been rated for difficulty. The ratings run from 1 (for the easiest) to 8 (for the most difficult)

1	C / 3		31	C / 6
2	C / 4		32	B / 7
3	C / 5		33	B / 5
4	B / 5		34	C / 2
5	D / 4		35	C / 6
6	B / 5		36	C / 7
7	C / 5		37	B / 6
8	B / 6		38	D / 6
9	D / 5		39	A / 4
10	B / 4		40	D / 6
11	B / 4		41	D / 6
12	C / 7		42	T / 8
13	A / 8		43	F / 5
14	D / 5		44	T / 5
15	D / 7		45	F / 6
16	B / 3		46	T / 4
17	C / 2		47	F / 8
18	B / 7		48	T / 6
19	D / 4		49	T / 3
20	B / 3		50	T / 5
21	B / 7		51	T / 2
22	C / 6		52	T / 6
23	C / 8		53	F / 7
24	B / 5		54	T / 7
25	B / 7		55	library / 7
26	B / 5		56	T / 5
27	D / 4		57	tomato / 6
28	A / 6		58	C / 5
29	C / 5		59	D / 3
30	C / 7		60	A / 3

Integrating Themes
Program 6
Microbial Evolution

> As is often the case, epoch-making ideas carry with them
> implicit, unanalyzed assumptions that ultimately impede
> scientific progress until they are recognized for what they
> are. So it is with the prokaryote-eukaryote distinction. Our
> failure to understand its true nature set the stage for the
> sudden shattering of the concept when a "third form of life"
> was discovered in the late 1970s, a discovery that actually left
> many biologists incredulous. Archaebacteria, as this
> third form has come to be known, have revolutionized our
> notion of the prokaryote, have altered and refined the way in
> which we think about the relationship between prokaryotes
> and eukaryotes. . .and will influence strongly the view we
> develop of the ancestor that gave rise to all extant life.
> – C. R. Woese and R. S. Wolfe

One of microbiology's fifty most significant events in the past 125 years:

1977

In this year, **Carl Woese** used ribosomal RNA analysis to identify a third form of life, the *Archaea* whose
genetic makeup is distinct from but related to both *Bacteria* and *Eucarya*

Student Assignments

- **Watch**   Video 6: Microbial Evolution
- **Read**   Chapter 19: Microbial taxonomy and
  Chapter 20: The *Archaea* in *Microbiology*,
  4th ed, Prescott/Harley/Klein, McGraw-Hill
- **Lab Experiments** from *Laboratory Exercises in Microbiology*, 4th ed., Harley/Prescott,
  McGraw-Hill.
  Exercise 40: Using *Bergey's Manual of Systematic Bacteriology* to
  Identify Bacteria
  Exercise 41: General Unknown

Lesson Overview

Video 6: *Molecular Evolution*
Recent genetic techniques have led to new theories of evolution and the relationships between organisms. Students examine this "evolution revolution" using molecular sequences to trace the phylogenetic relationships of microbial life. Both the big picture of microbial evolution and the methods necessary for determining molecular phylogenies are examined.

Chapter 19: Microbial Taxonomy
Microorganisms are tremendously diverse in size, shape, physiology, and lifestyle. This chapter introduces the general principles of microbial taxonomy and presents an overview of the current classification scheme. Subsequent chapters will examine the various groups of microorganisms in greater detail.

Chapter 20: The *Archaea*
This chapter summarizes the properties of a diverse group of organisms known as the archeae. These organisms are very different from the eubacteria and from the eukaryotes. This chapter describes some of the major characteristics associated with each of the major groups of archaea.

Learning Objectives

After successfully completing this unit, the student will be able to:

- discuss the rationale behind the science of taxonomy and discuss the three domains of living organisms (*Eubacteria, Archaea,* and *Eucaryota*)
- discuss the meaning of the word species and the basis for grouping organisms into species
- discuss the two ways (phylogenetic and phenetic) of classifying organisms
- discuss the various characteristics used in taxonomy and explain why nucleic acid sequences are probably the best indicators of microbial phylogeny and relatedness
- discuss the classification scheme(s) used in *Bergey's Manual of Systematic Bacteriology*
- discuss the dynamic nature of bacterial taxonomy and the new types of data that are contributing to the changes being made
- discuss the morphological and physiological diversity of the archaea
- discuss the difference between the cell walls of archaea and those of eubacteria
- describe the lipid composition of archaeal cell membranes
- discuss the general genetic, molecular, and metabolic characteristics of the archaea
- discuss the restricted habitats that are typical for the archaea
- describe the structural, chemical, and metabolic adaptations that allow archaea to grow in extreme environments

Course Integrator

Chapter 19

General Introduction and Overview (p. 394)
Web Link   *Microbiology* Online Learning Center
                                           www.mhhe.com/biosci/cellmicro/prescott/

Microbial Evolution and Diversity (p. 395)
Transparency   Universal Phylogenetic Tree Figure 19.3
Microbes in Motion II   Environmental Microbiology Book Diversity of
                                           Microorganisms Chapter/Archaea/p2
                                           Table 19.8 /pp 3,8
Video Unseen Life on Earth   Unit # 6 Microbial Evolution

Taxonomic Ranks (p. 397)

Classification System (p. 398)

Major Characteristics Used in Taxonomy (p. 400)

Assessing Microbial Phylogeny (p. 405)

The Major Divisions of Life (p. 407)
Microbes in Motion II   Environmental Microbiology/Diversity of Microorganism

Web Link   *Tree of Life*:
                                           http://phylogeny.arizona.edu/tree/phylogeny.html

Bergey's Manual of Systematic Bacteriology (p. 411)

A Survey of Bacterial Phylogeny and Diversity (p. 413)
Transparency   Phylogeny of the Achaea Figure 19.12
Transparency   Phylogeny of the Eubacteria Figure 19.13

Chapter 20

Introduction to the *Archaea* (p. 422)
Microbes in Motion II   Environmental Microbiology/Diversity of
                                           Microorganisms/Archaea
Web Link   Microbiology Online learning Center
                                           Http://www.mhhe.com/biosci/cellmicro/prescott/

Kingdom *Cranarchaeota* (p. 428)

Microbes in Motion II   Environmental Microbiology/Diversity of
                                            Microorganisms/Archaea

Kingdom *Euryarchaeota* (p. 429)
Transparency   Methane Synthesis Figure 20.13

Microbial Terms Used in Video 6 in Order of Appearance with Page Indexing to Textbook (*Microbiology*, 4th ed., Prescott/Harley/Klein, McGraw-Hill)

ameobae 821	nucleotides 203
DNA 212	protein synthesis 226-33
hydrogen gas 568	ribosome 228
hyperthermophiles 128	*Archaea 421-35*
Thermogota maritima 439	chromosome 325-26
genome 256	
ribosomal RNA 262	

Web Links Related to Video 6 (Molecular Evolution) and Readings in Chapter 19 (Microbial Taxonomy) and Chapter 20 (The *Archaea*) (Go to www.mhhe.com/prescott, Student On Line Learning Center, Double Click on Chapter 19 [and then 20], Double Click on Web Sites)

Chapter 19

The Tree of Life. The *Tree of Life* is a multi-authored, Internet project containing information about the diversity of organisms on Earth, their history, and characteristics. The information is linked together in the form of the evolutionary tree that connects all organisms to each other.
http://phylogeny.arizona.edu/tree/phylogeny.html

The Index Virum. The *Index Virum* presents lists of virus taxa that reflect the currently approved classification of the International Committee on Taxonomy of Viruses (ICTV).
http://life.anu.edu.au/viruses/Ictv/index.html

The international Committee on Taxonomy of Viruses (ICTV). From the National Center for Biotechnology.
http://www.ncbi.nlm.nih.gov/ICTV

Chapter 20

The Curious Microbes. Essays of the extremes and the unusual.
http://commtechlab.msu.edu/sites/dlc-me/curious/cindex.html

The Microbe Zoo.   This site is a virtual zoo filled with exotic specimens arranged in pavilions representing specific microbial habitats.
http://commtechlab.msu.edu/sites/dlc-me/zoo

Eukaryotes in Extreme Environments.   This site was compiled by Dave Roberts, Department of Zoology, the Natural History Museum, London.
http://www.nhm.ac.uk/zoology/extreme.html

Archaea in Space. An interesting site put together by NASA.
http://www.accessexcellence.org/WN/SU/arch998.html

From Primordial Soup to the Prebiotic Beach. An interview with exobiology pioneer, Dr. Stanley Miller, University of California San Diego.
http://www.accessexcellence.org/WN/NM/miller.html

Microbial Terms Used in Chapter 19 (Microbial Taxonomy)

Archaea 395
Bacteria 395
bacterial species 397
binomial system 398
biovar 397
classification 394
dendogram 400
domains 395
endosymbiotic 395
Eubacteria 395
Eucarya 395
evolutionary distance 406
genus 398

identification 394
Jaccard coefficient 399
melting temperature 402
molecular chronometers 405
morphovar 397
natural classification 398
nomenclature 398
nucleic acid hybridization 402
numerical taxonomy 398
oligonucleotide 406
phenetic system 398
phenons 400
phylogenetic tree 405

phylogeny 400
protists 409
serovar 397
similarity matrix 399
simple matching 399
species 397
strain 397
stromatolites 395
systematics 394
taxon 394
taxonomy 394
type strain 397

Microbial Terms Used in Chapter 20

Archaea 422
archaebacteria 422
bacteriorhodopsin

extreme halophiles 433
halobacteria 433
methanogens 429

pseudomurein 422
purple membrane 433
thermoacidophiles 428

Sample Test Questions from Video 6 (Molecular Evolution)

1. The theory of evolution holds that all organisms evolved from a common ancestor.
   True
   False

2. In the time clock presented, when did humans first appear? _______________________

3. The termal vents that Karl Stetter collected samples from were approximately _____ °C.
   a. 50
   b. 70
   c. 80
   d. 90

4. The microbiological term for the microbes that Karl Stetter collected is:
   a. thermophiles
   b. mesophiles
   c. psychrophiles
   d. hyperthermophiles

5. The microbes that Karl Stetter collected used _____ gas as their energy source.
   a. oxygen
   b. nitrogen
   c. hydrogen
   d. phosphorous

6. In the 1970's, Carl Woese proposed a _____-domain system for classification that is now widely accepted by microbiologists.
   a. two
   b. three
   c. four
   d. five

7. Organisms in a domain are further divided into __________.
   a. species
   b. orders
   c. families
   d. kingdoms

8. With the exception of viruses, all organisms have rRNA.
   True
   False

9. What percentage of a human's genes are the same as the genes of a whale, a dog, and a monkey?
   a. 50%
   b. 75%
   c. 90%
   d. 97%

10. Only a few microbes can thrive at hot temperatures.
    True
    False

Sample Test Questions from Text Chapter 19 (Microbial Taxonomy) Web Site

1. Taxonomy is the science of biological classification.
    O True
    O False

2. Taxonomy is composed of three parts: classification, nomenclature and identification.
    O True
    O False

3. Phylogenetic classification is based on
    O possible evolutionary relationships
    O overall similarity

4. Phenetic classification is base on
    O posible evolutionary relationships
    O overall similarity

5. Which of the following is true about *Bergey's Manual of Systematic Bacteriology*?
    O The first edition is mostly phenetic while the second edition is more phyletic.
    O The first edition is largely phyletic while the second edition is more phenetic.
    O The first and second editions are both phenetic.
    O The first and second editions are both largely phyletic.

6. Phenetic classification in *Bergey's Manual* is based on easily determined features such as
    O Gram stain
    O Cell shape
    O Motility
    O All of these

7. The general order of classification is
    O kingdom, phyla, order, class, family, genus, species
    O kingdom, phyla, order, family, class, genus, species
    O kingdom, phyla, class, family, order, genus, species
    O kingdom, phyla, class, order, family, genus, species

8. Which of the following are not characteristics used in taxonomic organization?
    O morphological
    O physiological
    O ecological
    O all of these are used.

9. A population descending from a single organism or pure culture isolate is called a
   O genus
   O strain
   O species
   O subspecies

10. The basic taxonomic group is the
    O species
    O genus
    O family

11. Species is defined in terms of
    O sexual reproduction
    O general similarity
    O both sexual reproduction and general similarity

12. A bacterial species is a collection of strains that have many stable properties in common.
    O true
    O false

13. Bacterial taxonomy is rapidly due to the acquisition of new data, particularly the use of molecular techniques such as
    O Comparison of ribosomal RNA structure
    O Chromosome sequences
    O Both of these

14. Which of the following is true about the G + C content of organisms?
    O G + C content of organisms is easily determined
    O G + C content is taxonomically valuable
    O G + C content indirectly reflects the base sequence of organisms
    O All of these are true about the G + C content of organisms.

15. The organisms in which of the following are more closely related?
    O order
    O family
    O class
    O phylum

16. The binomial system of nomenclature assigns each organism a scientific name consisting of
    O species and strain
    O class and species
    O genus and species
    O family and genus

17. Conjugation is useful in determining relatedness between two organisms because it generally does not cross genera.
    O true
    O false

18. Usually phylogenetic and phenetic classification schemes for bacteria agree closely with one another.
    O true
    O false

19. The earliest bacteria were probably:
    O aerobic
    O anaerobic
    O heterotrophic
    O photosynthetic

Sample Test Questions from Text Chapter 20 (The *Archaea*) Web Site

1. *Archaea* differ from bacterial groups in that they have special structural, chemical, and metabolic adaptions that enable them to grow in extreme environments.
    O true
    O false

2. *Archaea* usually grow in a few restricted or specialized habitats such a
    O anaerobic
    O hypersaline
    O extremely hot
    O extremely cold
    O all of these

3. Methanogenic bacteria are archaebacteria that are
    O strict anaerobes that obtain energy through the synthesis of methane.
    O strict anaerobes that obtain energy through the consumption of methane
    O strict aerobes that obtain energy through the synthesis of methans.

4. The archaebacteria lack which of the following that are normally found in gram-negative eubacteria?
    O outer membrane
    O a complex peptidoglycan network
    O they lack both outer membrane and a complex peptidoglycan network
    O they lack neither outer membrane nor a complex peptidoglycan network

5. Which of the following is currently the largest group of archaebacteria?
   O extreme halophiles
   O methanogens
   O extreme thermophiles
   O sulfate reducers

6. Methanogens may contribute to the greenhouse effect and global warming.
   O true
   O false

7. Methanogens are potentially of great importance because
   O they produce methane
   O they consume methane
   O methane is an excellent energy source
   O both a and c
   O both b and c

8. Some archaebacteria are symbionts in animal digestive tracts.
   O true
   O false

9. Recently, archaebacteria have been found in cold environments.
   O true
   O false

10. Archaeobacteria may constitute up to ______% of the procaryotic biomass in Antarctic surface waters.
    O 23
    O 28
    O 34
    O 43

11. Which of the following have been suggested to be among the first organisms to have evolved?
    O methanococcus
    O methanopyrales
    O methylococcus
    O none of the above is corrrect

12. Extreme halophiles are found in
    O animal intestines
    O salt lakes
    O the Antarctic ice mass

Chapter 19 Sample Test Questions on Microbial Taxonomy

1.   The arrangement of organisms into groups is best described as
        a. taxonomy.
        b. classification.
        c. systematics.
        d. identification.

2.   The science dealing with classification is called
        a. taxonomy.
        b. nomenclature.
        c. systematics.
        d. identification.

3.   The determination of the taxon to which an organism belongs is called
        a. taxonomy.
        b. classification.
        c. systematics.
        d. identification.

4.   The assignment of names to taxonomic groups is referred to as
        a. taxonomy.
        b. nomenclature.
        c. systematics.
        d. identification.

5.   The scientific study of organisms with the ultimate goal of characterizing and arranging them in an orderly manner is
        a. taxonomy.
        b. classification.
        c. systematics.
        d. identification.

6.   The hierarchy of classification is
        a. kingdom, phyla, order, class, family, genus, species.
        b. kingdom, phyla, order, family, class, genus, species.
        c. kingdom, phyla, class, family, order, genus, species.
        d. kingdom, phyla, class, order, family, genus, species.

7. A population descending from a single organism or pure culture isolate is called a
   a. genus.
   b. species.
   c. subspecies.
   d. strain.

8. Which of the following is an incorrect way of representing a particular organism with the binomial system of nomenclature?
   a. *Escherichia coli*
   b. *E. coli*
   c. *Esch. coli*
   d. Both *E. coli* and *Esch. coli* are incorrect.

9. A classification system based on mutual similarity comparing as many characteristics as possible is called a ___________ system.
   a. phylogenetic
   b. phyletic
   c. phenetic
   d. either phylogenetic or phyletic but not phenetic

10. A classification system based on evolutionary relationships is called a __________ system.
    a. phylogenetic
    b. phyletic
    c. phenetic
    d. either phylogenetic or phyletic but not phenetic

11. Numerical taxonomy has been feasible only with the advent of
    a. the compound microscope.
    b. the electron microscope.
    c. computers.
    d. the abacus.

12. In order to use numerical taxonomy, at least 50 __________ characteristics should be measured for comparison.
    a. morphological
    b. biochemical
    c. physiological
    d. All of these.

13. In numerical taxonomy, the __________ coefficient involves the proportion of common attributes either present or absent.
    a. simple matching
    b. associative
    c. Jaccard

14. In numerical taxonomy, the __________ coefficient involves the proportion of common attributes present but ignores characters that are missing from both organisms being compared.
    a. simple matching
    b. associative
    c. Jaccard
    d. taxonomial

15. When a numerical taxonomy scheme is used, groups of organisms with   great mutual similarity are called
    a. species.
    b. taxons.
    c. numerons.
    d. phenons.

16. Which of the following are not characteristics used in taxonomic organization?
    a. morphological
    b. physiological
    c. ecological
    d. All of these are used.

17. Which of the following is (are) a useful way(s) of comparing proteins for the determination of relatedness?
    a. amino acid sequences of proteins with similar functions
    b. immunologic cross reactivity
    c. enzyme properties
    d. All of these.

18. Nucleic acid base composition (% G + C) is useful for determining relatedness at the __________ level.
    a. kingdom
    b. class
    c. genus
    d. None of these.

19. Hydrogen bonding between DNAs from different organisms is referred to as
    a. hydrolysis.
    b. restriction.
    c. hybridization.
    d. ligation.

20. When using nucleic acid hybridization in determining relatedness, one must carefully control the __________ because this strongly affects the degree of sequence homology needed for stable hybrid formation.
    a. temperature
    b. humidity

21. Living organisms are currently classified into __________ kingdoms.
    a. two
    b. three
    c. four
    d. five

22. All procaryotic organisms are currently classified in the kingdom
    a. *Fungi*
    b. *Protista*
    c. *Monera*
    d. Some are found in each of the kingdoms listed.

23. *Bergey's Manual* currently classifies bacteria
    a. phylogenetically.
    b. phenetically.
    c. numerically.
    d. None of these.

24. The temperature at which the two strands of a double-stranded DNA molecule will separate from each other is called the __________ temperature.
    a. separation
    b. melting
    c. destabilizing
    d. mobilization

25. Which of the following has not been used in systematics?
    a. physiology
    b. epidemiology
    c. ecology
    d. All of these have been used in systematics.

26. Which of the following gene exchange mechanisms is not useful in classification studies?
    a. transformation
    b. transduction
    c. conjugation
    d. All of these are useful in classification studies.

27. Which of the following is not true about the G + C content of organisms?
    a. Organisms with similar G + C content have similar base sequences.
    b. Organisms with different G + C content have dissimilar base sequences.
    c. Similar G + C content only suggests relatedness if the two organisms are also alike phenotypically.
    d. All of these are true about the G + C content of organisms.

28. Photoautotrophic organisms are usually found in which of the following kingdoms?
    a. *Plantae*
    b. *Protista*
    c. *Monera*
    d. All of these.

29. Organisms with ingestive nutrition are usually found in which of the following kingdoms?
    a. *Animalia*
    b. *Protista*
    c. both *Animalia* and Protista
    d. neither *Animalia* nor *Protista*

30. The concept of a species as a group of interbreeding, or potentially interbreeding, populations reproductively isolated from other such groups is satisfactory for
    a. higher organisms.
    b. microorganisms.
    c. both higher organisms and microorganisms.
    d. neither higher organisms nor microorganisms.

31. Which of the following is more useful for distantly related organisms?
    a. DNA-DNA hybridization
    b. DNA-RNA hybridization using mRNA
    c. DNA-RNA hybridization using rRNA
    d. All of these are equally useful.

32. A treelike diagram summarizing the relatedness of organisms as determined by numerical taxonomy is called
    a. geneogram.
    b. dendrogram.
    c. taxogram.
    d. sonogram.

33. Organisms that are either multicellular or unicellular eucaryotes, have cell walls, and have absorptive nutrition are found in which of the following kingdoms?
    a. *Monera*
    b. *Plantae*
    c. *Fungi*

34. In which kingdom are found unicellular, photosynthetic, eucaryotic organisms?
     a. *Fungi*
     b. *Protista*
     c. *Monera*
     d.. All of these.

35. The organisms in which of the following are more closely related?
     a. order
     b. family
     c. class
     d. phylum

36. The binomial system of nomenclature assigns each organism a scientific name consisting of
     a. species and strain.
     b. kingdom and species.
     c. genus and species.
     d. kingdom and genus.

37. T F Numerical taxonomy is based primarily on the percent DNA sequence homology.

38. In the classification scheme proposed by Carl Woese, the term domain (or empire) would be used to describe:
     a. a new level above the kingdom.
     b. a new level below the kingdom.
     c. the same level currently called the kingdom.
     d. None of these (the term domain would not be used).

39.T F Phylogenetic studies involving 16S rRNA sequences include signature sequences that have been identified for eubacteria, archaebacteria, and eucaryotes.

40. Bacterial strains that are characterized by biochemical differences are called __________.

41. Bacterial strains that differ morphologically are called __________.

42. Bacterial strains that have distinct antigenic properties are called __________.

43. A general term used to describe groups based on mutual similarity or evolutionary relatedness is
     a. class.
     b. taxa.
     c. system.
     d. group.

44.T F In microorganisms as in higher organisms, species are usually defined by sexual interbreeding capabilities and by reproductive isolation.

45.T F Chromosomal gene exchange is not useful in classification studies because procaryotes do not reproduce sexually.

46.T F Microorganisms do not generally reproduce sexually, therefore, species are usually defined by phenotypic and genotypic similarities.

47.T F Microbial species are collections of strains that share many stable properties in common but differ significantly from other strains.

48.T F The type strain is a well-characterized strain to which other strains are compared for inclusion in or exclusion from a particular species.

49.T F The type strain is the most representative strain of a particular species.

50.T F Conjugation is useful in determining relatedness between two organisms because it generally does not cross genera.

51.T F Transformation is not useful in determining relatedness between two organisms because it frequently crosses genera.

52.T F In cases in which numerical taxonomy has been applied, it always agrees with existing classification schemes.

53.T F Phenons determined in a numerical taxonomic scheme usually parallel closely species determined in other schemes.

54.T F Usually phylogenetic and phenetic classification schemes for bacteria agree closely with one another.

55.T F Although there are other classification schemes, the one used in Bergey's Manual is currently the standard.

56. Any system that reflects the biological nature of organisms as much as possible and arranges them into groups whose members share many common characteristics is referred to as a __________ classification system.

57. Phylogenetic classification has recently included the comparison of 16S rRNA sequences, including specific oligonucleotide __________ sequences that have been identified for eubacteria, archaeobacteria, and eucaryotes.

58. Which of the following is true about *Bergey's Manual*?
    a. The first edition is mostly phenetic while the second edition is more phyletic.
    b. The first edition is largely phyletic while the second edition is more phenetic.
    c. The first and second editions are both largely phenetic.
    d. The first and second editions are both largely phyletic.

59. Layered or stratified rocks that are formed by incorporation of minerals into microbial mats are referred to as ________________.

60. The earliest bacteria were probably:
    a. aerobic
    b. anaerobic
    c. heterotrophic
    d. photosynthetic

61. The ________________ ________________ proposes that mitochondria and chloroplasts developed from free-living procaryotes that invaded a precursor to the eucaryotes and established a stable relationship.

Chapter 20 Sample Test Questions on the *Archaea*

1. Which of the following groups of the *Archaea* do not have cell walls?
    a. halobacteria
    b. thermoplasma
    c. thermococci
    d. None of the above have cell walls.

2. Which of the following contribute(s) to the differences between the archaebacteria and the eubacteria?
    a. archaebacteria lack muramic acid.
    b. archaebacteria have isopranyl glycerol ethers rather than fatty acid esters in their membrane lipids.
    c. archaebacteria differ from eubacteria in their tRNA composition, ribosome structure, antibiotic sensitivity, translation elongation factors, initiator tRNA, and DNA-dependent RNA polymerase structure.
    d. All of these.

3. Methanogenic bacteria are archaebacteria that are
    a. obligate anaerobes that produce methane.
    b. obligate anaerobes that consume methane.
    c. obligate aerobes that produce methane.
    d. obligate aerobes that consume methane.

4. Which of the following is not true of the extreme thermophilic archaebacteria in the kingdom of *Crenarchaeota*?
   a. They are sulfur dependent.
   b. They are acidophiles.
   c. Their cell walls have lipoprotein and carbohydrate but no peptidoglycan.
   d. All of these are true of the extreme thermophilic bacteria.

5. A peptidoglycan-like polymer with L-amino acids instead of D-amino acids is called __________.

6. A modified cell membrane that carries out photosynthesis using bacteriorhodopsin in the absence of chlorophyll is called a __________ membrane.
   a. colorless
   b. orange
   c. purple
   d. green

7. An organism with a temperature growth optimum near 80 °C and a pH growth optimum between 1.0 and 6.5 is usually referred to as a(n)
   a. halophile.
   b. thermoacidophile.
   c. thermophile.
   d. thermoalkalophile.

8. An organism that requires at least 1.5 M NaCl in order to grow is referred to as a(n)
   a. facultative halophile.
   b. extreme halophile.
   c. sodiophile.
   d. barophile.

9. Which of the following do methanogens not use to form methane?
   a. carbon dioxide
   b. formate
   c. acetate
   d. They use all of these to form methane.

10. T F The genome of *Methanococcus jannaschii* has been completely sequenced and has been found to be genotypically indistinct from the eubacteria.

11. The archaebacteria lack which of the following that are normally found in gram-negative eubacteria?
   a. outer membrane
   b. a complex peptidoglycan network
   c. they lack both outer membrane and a complex peptidoglycan network
   d. they lack neither outer membrane nor a complex peptidoglycan network

12. Archaebacterial membranes contain which of the following polar lipids?
    a. phospholipids
    b. sulfolipids
    c. glycolipids
    d. All of these.

13. Autotrophy is widespread among the
    a. methanogens.
    b. extreme thermophiles.
    c. both methanogens and extreme thermophiles.
    d. neither methanogens nor extreme thermophiles.

14. Autotrophic archaebacteria incorporate carbon dioxide
    a. by a single mechanism for all species.
    b. by various mechanisms between genera.
    c. by various mechanisms between species.
    d. There are no autotrophic archaebacteria.

15. Methanogens thrive in __________ environments.
    a. aerobic
    b. anaerobic
    c. microaerophilic
    d. All of these.

16. Methanogens thrive in an environment that is rich in
    a. organic material.
    b. inorganic material.
    c. sodium ions.
    d. chloride ions.

17. Archaebacteria normally do not use __________ as a mechanism for reproduction.
    a. binary fission
    b. budding
    c. mitosis
    d. fragmentation

18. T F The methanogens do not have a functional TCA cycle.

19. T F Extreme thermophiles do not have a functional TCA cycle.

20. The primary lipid components of the membranes of extreme thermophiles are
    a. sulfolipids.
    b. cholesterol.
    c. 40 carbon tetraethers.

21.T F Unlike other procaryotes polygenic mRNA has not been found in archaeobacteria.

22.T F The archaebacteria are like eucaryotes in that mRNA splicing has been observed.

23.   Which of the following is not true about archaebacterial ribosomes?
    a.   they are 70S like eubacterial ribosomes
    b.   their shape differs from both eubacterial and eucaryotic ribosomes
    c.   they have similar antibiotic sensitivities to those of eubacteria
    d.   All of these are true about archaebacterial ribosomes.

24.   Which of the following is currently the largest group of archaebacteria?
    a.   extreme halophiles
    b.   methanogens
    c.   extreme thermophiles
    d.   sulfate reducers

25.TF Methanogens may contribute to the greenhouse effect and global warming.

26.   Which of the following has been shown to cause corrosion in iron pipes?
    a.   extreme thermophiles
    b.   extreme halophiles
    c.   extreme ferrophiles
    d.   methanogens

27.   Methanogens are potentially of great importance because
    a.   they produce methane.
    b.   methane is an even-burning fuel.
    c.   methane is an excellent energy source.
    d.   All of these contribute to the potential importance of methanogens.

28.T F If the NaCl concentration drops below 1.5 M, the cell walls of *Halobacterium* become more rigid.

29.T F If the NaCl concentration drops below 1.5 M, the cell walls of *Halobacterium* disintegrate.

30.T F Some members of the archaebacteria lack a cell wall.

31.   Sulfolobus normally uses oxygen as an electron acceptor but it can use
    a.   sulfur.
    b.   ferric iron.
    c.   nitrogen.
    d.   All of these.

32. The cell membranes of wall-less archaebacteria can be strengthened by
    a. diglycerol ethers.
    b. glycoproteins.
    c. lipopolysaccharides.
    d. All of these.

33. Because archaebacteria lack peptidoglycan in their cell walls they are resistant to
    a. Beta-lactam antibiotics.
    b. lysozyme.
    c. both Beta-lactam antibiotics and lysozyme.
    d. neither Beta-lactam antibiotics nor lysozyme.

34. The chromosomal DNA of archaebacteria is normally __________ the chromosomal DNA of eubacteria.
    a. larger than
    b. smaller than
    c. the same size as
    d. It differs with different archaebacteria.

35. T F Archaebacterial sulfate reducers use elemental sulfur as an electron acceptor.

36. T F Some archaebacteria are symbionts in animal digestive tracts.

37. T F Archaebacterial membranes contain nonpolar lipids that are derivatives of squalene.

38. T F Many plasmids have been found in archaeobacteria.

39. T F The RNA polymerase enzymes of archaebacteria are more similar to eucaryotic enzymes than to eubacterial enzymes.

40. Which of the following is not an environment in which archaebacteria are normally found?
    a. animal intestinal tracts
    b. freshwater and marine sediments
    c. anaerobic protozoa
    d. Archaebacteria are found in all of these environments.

41. T F Picrophilus has a pH optimum below 1 and can even grow at or near pH 0.

42. T F Archaeglobales have been isolated from marine hydrothermal vents.

43. The purple membrane of *Halobacterium salinarum* contains
    a. bacteriorhodopsin
    b chlorophyll
    c both bacteriorhodopsin and chlorophyll
    d. neither bacteriorhodopsin nor chlorophyll

44.T F The DNA of the thermoacidophilic cocci is stabilized by association with histonelike proteins forming particles resembling eucaryotic nucleosomes.

45.T F Some members of the genus *Pyrodictum* can grow quite well above the   boiling point of water.

46.T F Recently, archaebacteria have been found in cold environments.

47. Archaebacteria may constitute up to __________% of the procaryotic biomass in Antarctic surface waters.
    a. 23
    b. 28
    c. 34
    d. 43

48. Which of the following is true about the cell-wall-less archaebacterial genus *Thermoplasma*?
    a   They are spherical at temperatures below 59°C
    b. They are filamentous at temperatures above 59°C
    c. Both a and b are correct.
    d. Neither a nor b is correct.

49. Geothermally heated water or soils that contain elemental sulfur are referred to as _______.

50. Which of the following have been suggested to be among the first living organisms to have developed?
    a. *Methanococcus*
    b. *Methanopyrales*
    c. *Methylococcus*
    d. None of the above is correct.

Answers to Sample Test Questions from Video 6

1. True
2. A few seconds before midnight.
3. D
4. D
5. C

6. B
7. D
8. A
9. D
10. False

Answers to Sample Test Questions from Text Chapter 19 (Microbial Taxonomy) Web Site

1. True
2. True
3. Possible evolutionary relationships.
4. Overall similarity.
5. The first edition is mostly phenetic while the second edition is more phyletic.
6. All of these.
7. Kingdom, phyla, class, order, family, genus, species.
8. All of these are used.
9. Strain
10. Species
11. Both sexual reproduction and general similarity.
12. True
13. Both of these.
14. All of these are true about the $G + C$ content of organisms.
15. Family
16. Genus and species
17. True
18. False
19. Anaerobic

Answers to Sample Test Questions from Text Chapter 20 Web Site

1. True
2. All of these.
3. Strictly anaerobes that obtain energy through the synthesis of methane
4. They lack both outer membrane and a complex peptidoglycan network.
5. Methanogens
6. True
7. Both a and c
8. True
9. True
10. 34

11. Methanopyrales
12. Salt lakes

Answers to Chapter 19 Sample Test Questions on Microbial Taxonomy

The question in this test bank have been rated for difficulty. The ratings run from 1 (for the easiest) to 8 (for the most difficult).

#	Answer / Diff.		#	Answer / Diff.
1	B / 2		28.	D/4
2	A / 5		29.	C/4
3	D / 4		30.	A/5
4	B / 2		31.	C/8
5	C / 6		32.	B/7
6	D / 7		33.	C/5
7	D / 5		34.	B/4
8	C / 6		35.	B/6
9	C / 5		36.	C/6
10	D / 7		37.	F/4
11	C / 5		38.	A/8
12	D / 5		39.	T/6
13	A / 7		40.	Biovars/6
14	C / 7		41.	Morphovars/6
15	D / 8		42.	Serovars/6
16	D / 3		43.	B/6
17	D / 5		44.	F/4
18	C / 7		45.	F/6
19	C / 4		46.	T/5
20	temperature / 6		47.	T/5
21	D / 1		48.	T/5
22	C / 3		49.	F/8
23	B / 1		50.	T/6
24	B / 5		51	F/6
25	D / 4		52.	F/4
26	B / 6		53.	F/6
27	A / 7		54.	F/6

55.	T/4
56.	Natural/7
57.	Signature/7
58.	A/3
59.	Stromatolites/7
60.	B/6
61.	Endosymbiotic hypothesis/6

Answers to Chapter 20 Sample Test Questions on the *Archaea*

The question in this test bank have been rated for difficulty. The ratings run from 1 (for the easiest) to 8 (for the most difficult).

#	Answer / Difff	#	Answer / Difff
1	B / 7	28.	F/6
2	D / 7	29.	T/6
3	A / 4	30.	T/3
4	D / 7	31.	B/8
5	pseudomurein	32.	D/5
6	C / 6	33.	C/7
7	B / 5	34.	B/7
8	B / 5	35.	F/6
9	D / 7	36.	T/5
10	F / 5	37.	T/5
11	C / 4	38.	F/6
12	D / 6	39.	T/7
13	C / 4	40.	D/6
14	C / 8	41.	T/6
15	B / 4	42.	T/4
16	A / 5	43.	A/6
17	C / 5	44.	T/7
18	T / 6	45.	T/5
19	F / 6	46.	T/5
20	C / 7	47.	C/7
21	F / 4	48.	C/7
22	F / 6	49.	Sofatara/7
23	C / 8	50.	B/8
24	B / 3		
25	T / 5		
26.	D/7		
27.	D/5		

Microbial Diversity and Evolution
Program 7
Microbial Diversity

> What's in a name? that which we call a rose
> By any other name would smell as sweet....
> —W. Shakespeare

One of microbiology's fifty most significant events in the past 125 years.

1986

In this year, **Kary Mullis** used a heat-stable enzyme from *Thermus aquaticus* to establish the polymerase chain reaction technology. The DNA polymerase replicates the DNA of interest plus oligonucleotide primers on either side. The primers are allowed to anneal and the reaction repeated to amplify the target DNA many-fold. Mullis was awarded the Nobel Prize in Chemistry in 1993.

Student Assignments

- **Watch**   Video 6: Microbial Diversity
- **Read**   Chapter 21: Bacteria: The Deinococci and Nonproteobacteria Gram Negatives
  Chapter 22: Bacteria: The Proteobacteria
  Chapter 23: Bacteria: The Low G + C Gram Positives
  Chapter 24: Bacteria: The High G + C Gram Positives in *Microbiology*, 4th ed., Prescott/Harley/Klein, McGraw-Hill.
- **Lab Experiments** from *Laboratory Exercises in Microbiology*, 4th ed., Harley/Prescott, McGraw-Hill.
  Exercise 14: Microbiological Culture Media Preparation and Sterilization
  Exercise 15: Culture Transfer Instruments and Techniques
  Exercise 16: Spread-Plate Technique
  Exercise 17: Streak-Plate Technique
  Exercise 18: Pour-Plate Techniques
  Exercise 19: Isolation of Pure Cultures and Their Maintenance
  Exercise 20: Cultivation of Anaerobic Bacteria

Lesson Overview

Video 7: *Microbial Diversity*
What is the relationship between bacteria, archeae, and eukaryote branches of the tree of life, with their startling variety of organisms? Students see comparisons of organisms in their natural habitats and examine ways of studying these organisms in those habitats and in the laboratory.

Chapter 21: Bacteria: The Deinococci and Nonproteinbacteria Gram Negatives
This chapter is devoted to nine of the more interesting and important eubacterial groups from volumes 1 and 5 of the 2nd edition of *Bergey's Manual of Systematic Bacteriology*. Though the organization and perspective of the 2nd edition is used, the description of each group in the current edition is summarized. Where appropriate the distinguishing characteristics, morphology, reproduction, physiology, metabolism, and ecology of each group is included. The taxonomy of each major group is summarized and representative species are discussed.

Chapter 22: Bacteria: The Proteobacteria
This chapter presents the diverse group of eubacteria known as the proteobacteria. The distinguishing characteristics of these gram-negative bacteria: morphology, physiology, metabolism, and ecology are presented. The phylogenetic relationships are discussed and representative species are examined.

Chapter 23: Bacteria: The Low G + C Gram Positives
This chapter describes the different approaches to Gram-positive organisms taken by the 1st and 2nd editions of *Bergey's Manual* and then focuses on the mycoplasmas, *Clostridium* and its relatives, and the bacilli and lactobacilli.

Chapter 24: Bacteria: The High G + C Gram Positives
This chapter surveys the general characteristics of the actinomycetes and other organisms that are classified as high G + C gram-positives in the 2nd edition of *Bergey's Manual*. The actnomycetes are filamentous bacteria that form branching hyphae and asexual spores.

Learning Objectives

After successfully completing this unit, the student will be able to:
- discuss the deeply branching eubacterial groups represented by the hyperthermophiles *Aquifex* and *Thermotoga*
- discuss the gram-positive deinococci
- discuss the photosynthetic bacteria that are described in volume 1 of the 2nd edition of *Bergey's Manual*
- discuss the chlamydiae and their loss of some of their metabolic independence, thereby becoming dependent on their host for their energy supply and for certain cellular constituents
- discuss gliding motility and its usefulness for organisms that digest insoluble nutrients or that move over the surfaces of solid substrata
- discuss the importance of this diverse group of organisms, either as disease agents or because of their effects on the habitat

- discuss the use of *Escherichia coli* as a major experimental organism that is studied in many laboratories
- describe the diverse life styles and metabolism of members of this group of organisms
- discuss the complex structures (prosthecae, stalks, buds, sheaths, or complex fruiting bodies produced by some members of this group
- discuss the ecological impact by chemolithotrophic bacteria
- discuss the dependence of parasitic eubacteria, such as *Bdellovibrio* and the rickettsia on their hosts for energy and/or cell constituents
- discuss the difference in classification of gram-positives that is used in the 1st and 2nd editions of *Bergey's Manual*
- discuss the variation in peptidoglycan structure that is useful in identifying specific groups
- discuss the various roles of these organisms: harmless, free-living saprophytes, pathogens, importance in the food and dairy industry
- describe the filamentous actinomycetes
- describe the morphology and arrangement of spores, cell wall chemistry, and the types of sugars present in cell extracts within the actinomycetes
- discuss the roles of actinomycetes in the mineralization of organic compounds and in the production of antibiotics
- describe the important human pathogens contained in the genera *Corynebacterium* and *Mycobacterium*
- describe the filamentous actinomycetes
- describe the morphology and arrangement of spores, cell wall chemistry, and the types of sugars present in cell extracts within the actinomycetes
- discuss the roles of actinomycetes in the mineralization of organic compounds and in the production of antibiotics
- describe the important human pathogens contained in the genera *Corynebacterium* and *Mycobacterium*

Course Integrator

Aquifex and the Termotogas (p. 439)

Microbes in Motion II	Environmental Microbiology/Diversity of Microorganisms /Archaea
Web Link	*Microbiology* Online Learning Center http://www.mhhe.com/biosci/cellmicro/prescott/

The Deinococci (p. 440)

Photosynthetic Bacteria (p. 440)

Visual Resource Library	Photosynthetic Pigments Figure 21.3
Visual Resource Library	Cyanobacterial Cell Structure Figure 21.8a
Transparency	Cyanobacterial Cell Structure Figure 21.8
Microbes in Motion II	Environmental Microbiology/Aquatic Habitats /Freshwater

The Planctomycetes (p. 449)

Visual Resource Library   The Chlamydial Life Cycle Figure 21.13
Transparency   The Chlamydial Life Cycle Figure 21.13
Microbes in Motion II   Miscellaneous Bacteria/Chlamydia

| *Microbes in Motion II* | Environmental Microbiology/Aquatic Habitats/Marine |

The δ-Proteobacteria (p. 481)

Visual Resource Library	Bdellovibrio Life Cycle Figure 22.32 a
Transparency	Bdellovibrio Life Cycle Figure 22.32 a
Transparency	Myxobacterial Life Cycle Figure 22.34
Visual Resource Library	Myxobacterial Life Cycle Figure 22.34
Visual Resource Library	Myxobacterial Fruiting Bodies Figure 22.35a
Microbes in Motion II	Environmental Microbiology/Terrestrial Habitats /Predatory Bacteria /Soil Microorganisms
Microbes in Motion II	Environmental Microbiology/Diversity of Microorganisms /Culture Independent Detection

The ∈-Proteobacteria (p. 485)

Chapter 23

Class Mollicutes (the Mycoplasmas) (p. 490)

| *Microbes in Motion II* | Miscellaneous Bacteria/Mycoplasma |
| Web Link | *Microbiology* Online Learning Center http://www.mhhe.com/biosci/cellmicro/prescott/ |

Low G+C Gram-Positive Bacteria in Bergeys Manual (p. 491)

| Visual Resource Library | Peptidoglycan Variation Figure 23.4 |
| *Microbes in Motion II* | Gram Positive Bacteria/Gram Positive Bacilli/Bacillus |

The Clostridia and Relatives (p. 493)

| *Microbes in Motion II* | Anaerobic Bacteria/Gram Positive Bacilli /Spore Forming Anaerobes |

The Bacilli and Lactobacilli (p. 495)

| *Microbes in Motion II* | Gram Positive Bacteria/Gram Positive Bacilli/Bacillus |
| *Microbes in Motion II* | Gram Positive Bacteria/Gram Positive Cocci: Streptococcus |

Chapter 24

General Properties of the Actinomycetes (p. 507)

Visual Resource Library	An Actinomycete Colony Figure 24.1
Visual Resource Library	Actinobacterial Classification Figure 24.3
Transparency	Actinobacterial Classification Figure 24.3
Web Link	*Microbiology* Online Learning Center http://www.mhhe.com/biosci/cellmicro/prescott/
Web Link	The Society for Actinomycetes http://www.nih.go.jp/saj/

Web Link   Actinomycetes-*Streptomyces* Internet Resource Center
http://www.cbs.umn.edu/asirc/

High C+G Gram-Positive Bacteria in *Bergey's Manual* (p. 509)

Microbes in Motion II   Anaerobic Bacteria/Gram Positive Bacilli
/Non-spore Forming Anaerobes
Microbes in Motion II   Gram Positive Bacteria/Gram Positive Bacilli
/Corynebacterium/
Microbes in Motion II   Miscellaneous Bacteria/Mycobacteria

Microbial Terms Used in Video 7 in Order of Appearance with Page Indexing to the Textbook (*Microbiology*, 4th ed., Prescott/Harley/ Klein, McGraw-Hill)

Thermogota	439	termite	567
gene amplification	314	shotgun sequencing	302
Proteobacteria	458	*Giardia*	554
Trichomonas	558	cloning	326

Web Links Related to Video 7 (Microbial Diversity) and Readings in Chapter 21 (Bacteria: The Deinococci and Nonproteobacteria Gram Negatives), Chapter 22 (Bacteria: The Proteobacteria), Chapter 23 (Bacteria: The Low G + C Gram Positives), and Chapter 24 (Bacteria: The High G + C Gram Positives) (Go to www.mhhe.com/prescott, Student on Line Learning Center, Double Click of Chapter 21[and then 22, 23, and 24], Double Click on Web Sites)

Chapters 21, 22, 23, 24

Cells Alive
http://www.cellsalive.com

Gram Stain A teaching page on the gram stain where you can view gram-stained images of bacteria that are medically important.
http://www.meddean.luc.edu/lumen/DeptWebs/microbio/med/gram/gram-stn.htm

Biological Identity of Procaryotes Keneth Todar, University of Wisconsin-Madison, Department of Bacteriology
http://www.bact.wisc.edu/Bact303/TheProcaryotes

Microbial Terms Used in Chapter 21 (Bacteria: The Deinococci and Nonproteobacteria Gram Negatives)

akinetes 445
anoxygenic 440
axial fibrils 450
baeocytes 448
chlamlydiae 449
chlorosomes 443
cyanobacteria 443
elementary body 449
gliding motility 454
green nonsulfur
  bacteria 442
green sulfur bacteria
  443
heterocysts 445
hormogonia 445
initial body 449
oxygenic
photosynthesis 440
phycobilisomes 444
trichome 445

Microbial Terms Used in Chapter 22 (Bacteria: The Proteobacteria)

α-proteobacter 459
β-proteobacter 466
binary fission 461
bioluminesence 476
budding 461
colorless sulfur
  bacteria 469
δ-proteobacter 481
enteric bacteria 477
enteritis 485
ε-proteobacter 485
fruiting body 484
γ-proteobacter 470
holdfast 463
methylotroph 462
mineralization
  process 475
myxobacteria 482
myxospores 484
nitrification 466
nitrifying bacteria
  466
prostheca 461
proteobacteria 459
purple nonsulfur
  bacteria 460
purple sulfur bacteria
  472
septicemia 485
sheath 468
stalk 461

Microbial Terms Used in Chapter 23 (Bacteria: The Low G + C Gram Positives)

coagulase 503
α-hemolysis 502
β-hemolysis 502
heterolactic
fermentation 499
lactic acid bacteria
  499
lancefield grouping
system 502
mycoplasmas 490
parasporal body 497

Microbial Terms Used in Chapter 24 (Bacteria: The High G + C Gram Positives)

acid-fast 513
actinobacter 510
actinomycete 507
actinomycetoma 516
conidia 507

conidospores 507
geosmin 516
madurose 518
mycolic acids 513
nocardioforms 514

snapping division 512
sporangiospores 509
streptomycetes 516
thallus 507

Sample Test Questions from Video 7 (Microbial Diversity)

1. There is really no environment on Earth that is not inhabited by some microbe.
   a. True
   b. False

2. Microbes are the oldest forms of life on Earth.
   a. True
   b. False

3. Each combination of a unique structure and a unique metabolism allows a microbe to exploit the resources of a particular environment.
   a. True
   b. False

4. Mutations are the basis for the diversity of life on Earth.
   a. True
   b. False

5. Most species of microbes has the same unique genetic blueprint.
   a. True
   b. False

6. Genetic diversity follows from metabolic diversity.
   a. True
   b. False

7. Most viruses have a common genetic code.
   a. True
   b. False

8. Making copies of strands of DNA is called __________.

9. The biosphere is fundamentally chemistry.
   a. True
   b. False

10. The biosphere is fundamentally chemistry.
    a. True
    b. False

11. The microbial world is the basis upon which our whole ecosphere rests.
    a. True
    b. False

Sample Test Questions from Text Chapter 21 (Bacteria: The Deinococci and Nonproteobacteria Gram Negatives)

1. The deinococci are gram-positive cocci and rods that are distinctive in their unusually great resistance to
   O radiation.
   O desiccation.
   O radiation and desiccation.
   O extreme temperatures

2. Which of the following is not a group of photosynthetic bacteria?
   O the green bacteria
   O the purple bacteria
   O the cyanobacteria
   O All of these are groups of photosynthetic bacteria.

3. The infectious stage of chlamydiae is called a(n) __________ body.
   O elementary
   O reticulate
   O contagious
   O oogonial

4. Cyanobacteria are procaryotes but carry out oxygenic photosynthesis like eucaryotes in that they:
   O use water as their electron source.
   O produce oxygen as a by-product of photosynthesis.
   O Both of these answers are correct.
   O Neither of these answers is correct.

5. Cyanobacteria have
    O a gram-negative cell wall.
    O a gram-positive cell wall.
    O an unique cell wall that is neither gram positive nor gram negative.
    O no cell walls.

6. Some eubacterial groups, such as the ____________ are deeply branching and very old.
    O hyperthermophiles, *Thermotogoa* and *Aquifex*
    O Cyanobacteria
    O Spirochetes

7. The ability of the deinococci to resist radiation is due to an unusual ability to repair chromosome damage, even fragmentation.
    O True
    O False

8. *Bacteroides* constitute as much as ______% of the bacteria isolated from human feces.
    O 10%
    O 20%
    O 30%
    O 40%

9. Cyanobacteria are so named because most species have a blue-green appearance caused by the photosynthetic pigment phycocyanin.
    O True
    O False

10. Some species of cyanobacteria can fix atmospheric nitrogen.
    O True
    O False

11. Cyanobacteria are frequently the photosynthetic partner in ________ associations.
    O mycorrhizal
    O rhizobial
    O lichen

12. Cytophages contribute significantly to wastewater treatment.
    O True
    O False

13. Some green sulfur bacteria are nonmotile but can control their depth by using gas vesicles to control buoyancy.
    O True
    O False

14. The spirochetes include the causative agents for
  - O syphilis
  - O Lyme disease
  - O both syphilis and Lyme disease.
  - O neither syphilis nor Lyme disease.

15. The bacteriochlorophyll pigments of purple and green bacteria enable them to live in
  - O saline environments
  - O deeper, anaerobic zones of aquatic habitats.
  - O shallow, warm aquatic habitats

Sample Test Questions from Text Chapter 22 (Bacteria: The Proteobacteria) Web Site

1. Proteobacteria are gram-negative bacteria of considerable importance as
  - O disease agents
  - O nitrogen fixers
  - O major experimental organisms
  - O all of these

2. Chemolithotrophic bacteria obtain energy and electrons by oxidizing inorganic nutrients rather than the organic nutrients employed by most bacteria.
  - O True
  - O False

3. *Bdellovibrio* and the rickettsias specialize in predatory or parasitic modes of existence.
  - O True
  - O False

4. Which of the following genera grow symbiotically within root nodules of legumes as nitrogen-fixing bacteroids?
  - O *Agrobacterium*
  - O *Rhizobium*
  - O *Pseudomonas*
  - O *Bacteroides*

5. Mineralization refers to
  - O the breakdown of organic materials to inorganic materials.
  - O the release of various minerals from ores.
  - O the utilization of minerals as energy sources.
  - O the incorporation of inorganic material into organic materials.

6. Pseudomonads are a large group of bacteria which can
   - O spoil refrigerated food
   - O cause disease
   - O photosynthesize
   - O a and b
   - O a and c

7. The genus *Pseudomonas* contains more than 70 species.
   - O True
   - O False

8. Pseudomonads and enteric bacteria are important experimental bacteria.
   - O True
   - O False

9. *Rickettsia rickettsii* is responsible for
   - O Q-fever
   - O Rocky Mountain spotted fever
   - O Lyme disease
   - O Typhoid fever

10. *Helicobacter pylori* is responsible for
    - O gastroenteritis.
    - O cholera.
    - O bacterial dysentery.
    - O peptic ulcer disease.

11. Whooping cough is caused by
    - O *Klebsiella pneumoniae*
    - O *Vibrio parahaemolyticus*
    - O *Bordetella pertussis*
    - O *Yersinia pestis*

12. In which of the following regions in the host bacterium are you most likely to find *Bdellovibrio*?
    - O the cytoplasm
    - O the nucleoid
    - O the space between the plasma membrane and the cell wall
    - O the space within the flagella

13. *Hyphomicrobium* may constitute as much as ____________ of the total bacterial population in oligotrophic (nutrient poor) freshwater habitats.
    - O 10%
    - O 25%
    - O 40%
    - O 60%

14. Bacteria that are capable of bioluminescence have been found in which of the following genera?
- O *Vibrio*
- O *Photobacterium*
- O *Vibrio* and *Photobacterium*
- O neither *Vibrio* nor *Photobacterium*

15. *Escherichia coli* is a good indicator of fecal contamination of water supplies because it is easily detectable.
- O True
- O False

16. *Escherichia coli* has not been found to be pathogenic in humans.
- O True
- O False

17. *Rickettsia* do not use glucose as an energy source.
- O True
- O False

18. Which of the following is true of *Agrobacterium tumefaciens*?
- O It is not capable of nitrogen fixation.
- O It causes crown gall disease when it carries a tumor inducing (Ti) plasmid.
- O It has been used to introduce foreign DNA into plant cells.
- O All of the above are true of *Agrobacterium tumefaciens*.

19. Which of the following genera fix nitrogen non symbiotically?
- O *Rhizobium*
- O *Nitrosomonas*
- O *Nitrobacter*
- O *Azotobacter*

20. The causative agent for cholera in humans is a member of the genus
- O *Escherichia*
- O *Vibrio*
- O *Enterobacter*
- O *Shigella*

Sample test Questions for Text Chapter 23 (Bacteria: The Low G + Gram Positives) Web Site

1. Proteobacteria are gram-negative bacteria of considerable importance as
- O disease agents
- O nitrogen fixers
- O major experimental organisms
- O all of these

2. Chemolithotropic bacteria obtain energy and electrons by oxidizing inorganic nutrients rather than the organic nutrients employed by most bacteria.
   - O   True
   - O   False

3. *Bdellovibrio* and the rickettsias specialize in predatory or parasitic modes of existence.
   - O   True
   - O   False

4. Which of the following genera grow symbiotically within root nodules of legumes as nitrogen-fixing bacteroids?
   - O   *Agrobacterium*
   - O   *Rhizobium*
   - O   *Pseudomonas*
   - O   *Bacteroides*

5. Mineralization refers to
   - O   the breakdown of organic materials to inorganic materials.
   - O   the release of various minerals from ores.
   - O   the utilization of minerals as energy sources.
   - O   the incorporation of inorganic material into organic materials.

6. Pseudomonads are a large group of bacteria which can
   - O   spoil refrigerated food
   - O   cause disease
   - O   photosynthesize
   - O   a and b
   - O   a and c

7. The genus *Pseudomonas* contains more than 70 species.
   - O   True
   - O   False

8. Pseudomonads and enteric bacteria are important experimental bacteria.
   - O   True
   - O   False

9. *Rickettsia rickettsii* is responsible for
   - O   Q-fever
   - O   Rocky Mountain spotted fever
   - O   Lyme disease
   - O   Typhoid fever

10. *Helicobacter pylori* is responsible for
    O   gastroenteritis.
    O   cholera.
    O   bacterial dysentery.
    O   peptic ulcer disease.

11. Whooping cough is caused by
    O   *Klebsiella pneumoniae*
    O   *Vibrio parahaemolyticus*
    O   *Bordetella pertussis*
    O   *Yersinia pestis*

12. In which of the following regions in the host bacterium are you most likely to find
    Bdellovibrio?
    O   the cytoplasm
    O   the nucleoid
    O   the space between the plasma membrane and the cell wall
    O   the space within the flagella

13. *Hyphomicrobium* may constitute as much as ______________ of the total bacterial population in
    oligotrophic (nutrient poor) freshwater habitats.
    O   10%
    O   25%
    O   40%
    O   60%

14. Bacteria that are capable of bioluminescence have been found in which of the following
    genera?
    O   *Vibrio*
    O   *Photobacterium*
    O   *Vibrio* and *Photobacterium*
    O   neither *Vibrio* nor *Photobacterium*

15. *Escherichia coli* is a good indicator of fecal contamination of water supplies because it is
    easily detectable.
    O   True
    O   False

16. *Escherichia coli* has not been found to be pathogenic in humans.
    O   True
    O   False

17. *Rickettsia* do not use glucose as an energy source.
    O   True
    O   False

18. Which of the following is true of *Agrobacterium tumefaciens*?
    O It is not capable of nitrogen fixation.
    O It causes crown gall disease when it carries a tumor inducing (Ti) plasmid.
    O It has been used to introduce foreign DNA into plant cells.
    O All of the above are true of Agrobacterium tumefaciens.

19. Which of the following genera fix nitrogen non symbiotically?
    O *Rhizobium*
    O *Nitrosomonas*
    O *Nitrobacter*
    O *Azotobacter*

20. The causative agent for cholera in humans is a member of the genus
    O *Escherichia*
    O *Vibrio*
    O *Enterobacter*
    O *Shigella*

Sample Test Questions from Text Chapter 24 (Bacteria: The High G +C Gram Positives)

1. Actinomycetes are aerobic, gram-positive bacteria that form branching hyphae (mycelia).
    O True
    O False

2. Actinomycetes are usually found in __________ habitats.
    O soil
    O freshwater
    O marine
    O all of these.

3. Actinomycetes form
    O substrate mycelia.
    O aerial mycelia.
    O both substrate mycelia and aerial mycelia.

4. Actinomycetes are the primary source of antibiotics we use in medicine.
    O True
    O False

5. Actinomycetes form spores that are
   - O   asexual
   - O   are used for sexual reproduction
   - O   both asexual and sexual
   - O   they do not form spores

6. The genera *Corynebacterium* and *Mycobacterium* contain important human pathogens.
   - O   True
   - O   False

7. The streptomycetes represent __________ of the viable organisms in the soil.
   - O   5%
   - O   10%
   - O   1-20%
   - O   1-10%

8. Of the nearly 500 species in the genus *Streptomyces*, only *S. somaliensis* is known to be pathogenic in humans.
   - O   True
   - O   False

9. Diphtheria is caused by a member of which of the following genera?
   - O   *Mycobacterium*
   - O   *Corynebacterium*
   - O   *Streptomyces*

10. Mycoplasmas are the smallest bacteria capable of self-reproduction.
   - O   True
   - O   False

11. Actinomycetes that dwell in the soil play an important role in plant and animal decomposition.
   - O   True
   - O   False

Chapter 21 Sample Test Questions on the Bacteria: Deinococci and Nonproteobacteria Gram Negatives

1. Which of the following is not a group of photosynthetic bacteria?
    a. the green bacteria
    b. the purple bacteria
    c. the cyanobacteria
    d. All of these are groups of photosynthetic bacteria

2.   Which of the following groups of photosynthetic bacteria are capable of oxygenic photosynthesis?
   a.   the green bacteria
   b.   the purple bacteria
   c.   the cyanobacteria
   d.   All of these.

3.   Anoxygenic photosynthetic bacteria use ___________ as their electron source.
   a.   hydrogen sulfide
   b.   sulfur
   c.   hydrogen
   d.   All of these.

4.   Which of the following is true about anoxygenic photosynthesis?
   a.   does not produce oxygen
   b.   often produces sulfur granules
   c.   Both of these answers are correct.
   D.   Neither of these answers is correct.

5.   Spherical cells of cyanobacteria that are able to fix atmospheric nitrogen are referred to as:
   a.   heterocysts
   b.   microcysts
   c.   oocysts
   d.   nitrocysts

6.   Spherical resting cells produced by *Sporocytophaga* are called:
   a.   heterocysts
   b.   microcysts
   c.   oocysts
   d.   nitrocysts

7.   The part of the spirochete that houses the cytoplasm and the nucleoid of spirochetes is called the _____________ cylinder.
   a.   ectoplasmic
   b.   cytoplasmic
   c.   protoplasmic
   d.   mesoplasmic

8.   The complex of periplasmic flagella that mediates the movement of spriochetes is referred to as the:
   a.   flagellar complex
   b.   axillary filament
   c.   axial filament
   d.   peritrichal axon

9. The infectious stage of chlamydiae is called a(n) _________ body.
   a. elementary
   b. reticulate
   c. contagious
   d. oogonial

10. The green sulfur bacteria
   a.. deposit granules within the cell.
   b. deposit granules outside the cell.
   c. deposit granules within and outside the cell.
   d. do not deposit granules.

11. Which of the following statements best describes the photosynthetic membranes of the green sulfur bacteria?
   a. They are unconnected to the plasma membrane.
   b. They are continuous with the plasma membrane.
   c.. They are attached to the plasma membrane but not continuous with it.
   d.. None of these.

12. The photosynthetic membranes of the green sulfur bacteria are called
   a. chlorosomes.
   b. chloroplasts.
   c. chlorocytes.
   d.. chlorophylls.

13. The green sulfur bacteria are
   a obligate anaerobes.
   b. facultative anaerobes.
   c. either obligate anaerobes or facultative anaerobes.
   d. neither obligate anaerobes nor facultative anaerobes.

14. Cyanobacteria are procaryotes but carry out oxygenic photosynthesis like eucaryotes; this means that
   a. they usewater as their electron source.
   b. produce oxygen as a by-product of photosynthesis.
   c. Both of these answers are correct.
   d. Neither of these answers is correct.

15. Although similar to eucaryotic photosynthetic organisms, cyanobacteria are different because they
   a. have only photosystem I not photosystem II.
   b. have only photosystem II not photosystem I.
   c. do not have chloroplasts.
   d. do not have any photosynthetic membranes.

16. Cyanobacteria are best described as
    a   photolithoautotrophs.
    b. chemoheterotrophs.
    c. photolithoautotrophs but can function as chemoheterotrophs if grown in the dark
    d. None of these.

17. Which of the following best describes cyanobacteria?
    a   They only carry out anoxygenic photosynthesis.
    b. They only carry out oxygenic photosynthesis.
    c. They carry out oxygenic photosynthesis if in an anaerobic environment.
    d. All carry out oxygenic photosynthesis, but some species can carry out anoxygenic
        photosynthesis if in an anaerobic environment

18. Cyanobacteria have
    a. a gram-negative cell wall.
    b. a gram-positive cell wall.
    c. an unique cell wall that is neither gram positive nor gram negative.
    d. no cell walls.

19. Some marine species of cyanobacteria
    a. have flagella and are, therfore, motile
    b. do not have flagella and are, therefore, nonmotil
    c. do not have flagella but are, nonetheless, motil by some unknown mechanism.
    d. do not have flagella but are motile by merans of pseudopods.

20 *Bacteroides* inhabit the intestinal tract of mammals and benefit the host by degrading which of the
    following?
    a. cellulose
    b. pectin
    c. comple carbohydrates
    d. all of the above are correct.

21. Which of the following accumulates sulfur granules outside the cell?
    a. purple sulfur bacteria
    b. green sulfur bacteria
    c. cyanobacteria
    d. none of the above

22. Which of the following is the oldest branch of eubacteria?
    a. *Thermotogoa*
    b. *Aquifex*
    c. Cyanobacteria
    d. Spirochetes

23. *Thermotogoa* generally grow in:
   a. marine hydrothermal vents.
   b. terrestrial solfataric hot spirngs.
   c. Both a and b are correct
   d. Neither a nor b is correct.

24. Which of the following is extremely radiation resistant?
   a. *Deinococcus*
   b. Aquifex
   c. Both a and b are correct.
   d. Neither a nor b is correct.

25. Which of the following contain both chlorophyll a and chlorophyll b?
   a. Cyanobacteria
   b. Prochlorales
   c. Both a and b are correct.
   d. Neither a nor b is correct.

26. In which of the following two genera of procaryotes are membrane-bound nuclear regions observed?
   a. *Gemmata*
   b. *Pirullela*
   c. Both a and b are correct
   d. Neitehr a nor b is correct.

27.TF The deinococci stain gram-positive but have a layered cell wall and an outer membrane that is more like a gram-negative organism.

28.TF The ability of the deinococci to resist radiation is due to an unusual ability to repair chromosome damage, even fragmentation.

29.TF The chlorosomes of chlorobia are continuous with the plasma membrane.

30.TF Chlamydiae are incapable of producing their own ATP and must rely on their host to supply it.

31.TF Although the function of the fleible outer sheath in which the axial filaments of spirochetes lay is unknown, it is essential (i.e. the bacteria will not survive if it is removed).

32. Dormant, thick-walled resting cells of cyanobacteria that are resistant to desiccation are called
   a. baeocytes.
   b. akinetes.
   c. hormogonia.
   d. thylakoids.

33. Small motile filaments of cyanobacteria are called
    a. baeocytes.
    b. oogonia.
    c. hormogonia.
    d. mycelia.

34. Bacteroides constitute as much as ______% of the bacteria isolated from human feces.
    a. 10%
    b. 20%
    c. 30%
    d. 40%

35. The purple non sulfur bacteria are
    a. obligate aerobes.
    b. facultative anaerobes.
    c. either obligate aerobes or facultative anaerobes
    d. neither obligate aerobes nor facultative anaerobes.

36. Cyanobacteria form __________, rows of bacterial cells that are in close contact with one another over a large area.
    a. hypha
    b. mycelia
    c. trichromes
    d. cell mats

37. Prochlorales lack which of the following?
    a. chlorophyll a
    b. chlorophyll b
    c. phycobilins
    d. They lack all of these.

38. T F Many bacteria with gliding motility are able to use as a nutrient source insoluble material that they encounter while gliding.

39. T F Cyanobacteria are so named because most species have a blue-green appearance caused by the photosynthetic pigment phycocyanin

40. T F Some species of cyanobacteria can fix atmospheric nitrogen.

41 T F Cyanobacteria are frequently the photosynthetic partner in lichen associations.

42. T F Cytophages contribute significantly to wastewater treatment.

43. *Aquifex* cannot use which of the following as electron donors?
    a. hydrogen
    b. thiosulfate
    c. sulfur
    d. oxygen

44. *Thermotoga* can grow anaerobically on which of the following?
    a. carbohydrates
    b. protein digets
    c. Both a and b are correct.
    d. Neither a nor b is correct.

45. The deinococci are resistant to:
    a. radiation.
    b. desiccation.
    c. radiation and desiccation.
    d. neither radiation nor desiccation.

46.T F Chloroflexis is not closely related to any eubacterial group.

47. Which of the following best describes the photosynthetic membranes of chlorobia?
    a. Accessory bacteriochlorophylls are located in the chlorosomes but the reaction center bacteriochlorophyll is located in the plasma membrane.
    b. Accessory bacteriochlorophylls are located in the plasma membrane but the reaction center bacteriochlorophyll is located in the chlorosomes.
    c. Accessory and reaction center bacteriochlorophylls are located in the chlorosomes.
    d. Accessory and reaction center bacteriochlorophylls are located in the plasma membrane.

48 T F Green sulfur bacteria are nonmotile but can control their depth by using gas vesicles to control buoyancy.

49. Which of the following are used by cyanobacteria for reproduction?
    a. binary fission
    b. budding
    c. fragmentation
    d. All of the above are correct.

50. Some cyanobacteria are not blue-green, rather they are red or brown; this is due to the presence of the photosynthetic pigment ___________.

51. Which of the following genera of the Planctomycetales have a stalk as part of their mechanism for attachment to surfaces?
    a. *Gemmata*
    b. Priullela
    c. *Planctomyces*
    d. All of the above.

52. The spirochetes include the casuative agents for:
    a. syphillis
    b. Lyme disease
    c. both syphillis and Lyme disease.
    d. neither syphillis nor Lyme disease.

53. The photosynthetic membranes of cyanobacteria are pairs of organelles called ________.

54. Which of the following genera of the *Sphingobacterica* are unable to degrade complex carbohydrates?
    a. *Cytophaga*
    b. *Sporocytophaga*
    c. *Flexibacter*
    d. All of the above are able to degrade complex carbohydrates.

55. T FA trichome is a bacterial cell with three different photosynthetic pigments.

Chapter 22 Sample Test Questions on Bacteria: The Proteobacteria

1. An extension of the cell that includes the plasma membrane and cell wall but that is narrower than the cell is referred to as a
   a. stalk.
   b. prostheca.
   c. stem.
   d. mycelium.

2. A nonliving appendage produced by a cell and extending from it is referred to as a
   a. stalk.
   b. prostheca.
   c. stem.
   d. mycelium.

3. A ___________ is a hollow tubelike structure that surrounds a chain of cells.

4. An organism that uses methane or methanol as its sole carbon and energy source is called a
   a. methanogen.
   b. methylotroph
   c. methanotroph
   d. methylogen

5. Members of the order *Desulfovibrionales* use which of the following as electron acceptors during anaerobic respiration.
   a. elemental sulfur
   b. sulfur compounds
   c. either elemental sulfur or sulfur compounds
   d. neither elemental sulfur nor sulfur compounds

6. *Rickettsia rickettsii* is responsible for
   a. Q-fever
   b. Rocky Mountain spotted fever
   c. Lyme disease
   d. Typhoid fever

7. *Helicobacter pylori* is responsible for
   a. gastroenteritis.
   b. cholera.
   c. bacterial dysentery.
   d. peptic ulcer disease.

8. Digestion of other bacteria and yeasts by myxobacteria is aided by its ___________ lifestyle.

9. Whooping cough is caused by
   a. *Klebsiella pneumoniae*
   b. *Vibrio parahaemolyticus*
   c. *Bordetella pertussis*
   d. *Yersinia pestis*

10. In the genus *Caulobacter* asymmetric transverse fission produces a small ________ cell that swims off.

11. *Hyphomicrobium* are chemoheterotrophic, aerobic, budding bacteria that frequently attach to solid objects in which of the following types of environments?
    a. marine
    b. freshwater
    c. terrestrial
    d. marine, freshwater, and terrestrial

12. Which of the following escape(s) from the phagosome before reproducing by binary fission?
    a. *Rickettsia*
    b. *Coxiella*
    c. *Rickettsia* and *Coxiella*
    d. neither *Rickettsia* nor *Coxiella*

13. Which of the following remain(s) in the phagosome after fusion with a lysosome and reproduces within the resulting structure?
a. *Rickettsia*
b. *Coxiella*
c. *Rickettsia* and *Coxiella*
d. neither *Rickettsia* nor *Coxiella*

14. In which of the following regions in the host bacterium are you most likely to find *Bdellovibrio?*
a. the cytoplasm
b. the nucleoid
c. the space between the plasma membrane and the cell wall
d. the space within the flagella

15. Q-fever is caused by
a. *Campylobacter jejuni*
b. *Coxiella burnettii*
c. *Helicobacter pylori*
d. *Bordetella pertussis*

16. Members of the genus *Pseudomonas* do which of the following?
a. mineralization of organic compounds
b. spoilage of refrigerated milk
c. Both a and b are correct.
d. Neither a nor b is correct

17. Mineralization refers to
a. the breakdown of organic materials to inorganic materials.
b. the release of various minerals from ores.
c. the utilization of minerals as energy sources.
d. the incorporation of inorganic material into organic materials.

18. *Hyphomicrobium* may constitute as much as ______________ of the total bacterial population in oligotrophic (nutrient poor) freshwater habitats.
a. 10%
b. 25%
c. 40%
d. 60%

19. The process of conversion of ammonia to nitrate is referred to as
a. nitrogen fixation.
b. nitrification.
c. ammonification.
d. denitrification.

20. Bacteria that are capable of bioluminescence have been found in which of the following genera?
a. *Vibrio*
b. *Photobacteria*
c. *Vibrio* and *Photobacteria*
d. neither *Vibrio* nor *Photobacteria*

21.TF *Escherichia coli* is a good indicator of fecal contamination of water supplies because it is easily detectable.

22. Myxobacteria lyse bacteria and yeasts by secretion of digestive enzymes. They then use the resulting peptides as a source of which of the following?
a. carbon
b. nitrogen
c. energy
d. carbon, nitrogen, and energy

23.TF The enterobacteria in the order *Enterobacteriales* are readily distinguishable from one another by morphological criteria.

24. The sheath of the *Comamonadaceae* performs which of the following functions?
a. helps attach bacteria to surfaces
b. helps obtain nutrients from slowly running water as it flows past
c. helps protect against predators
d. All of the above are correct.

25. Most of the oligotrophic proteobacteria are found in the
a. alpha proteobacteria.
b. beta proteobacteria.
c. gamma proteobacteria.
d. delta proteobacteria.

26. The largest group of proteobacteria are the
a. alpha proteobacteria.
b. beta proteobacteria.
c. gamma proteobacteria.
d. delta proteobacteria.

27. Which of the following best describes the photosynthetic membranes of the purple nonsulfur bacteria?
a. The photosynthetic membranes are attached to the plasma membrane but not continuous with it.
b. The photosynthetic membranes are not associated with the plasma membrane.
c. The photosynthetic membranes are continuous with the plasma membrane.
d. There are no photosynthetic membranes in the purple nonsulfur bacteria.

28.TF *Rickettsia* do not use glucose as an energy source.

29. The prostheca of *Caulobacter* is different from that of *Hyphomicrobium* in that it lacks cytoplasmic components and is composed almost totally of plasma membrane and cell wall.

30. Which of the following genera grow symbiotically within root nodules of legumes as nitrogen-fixing bacteroids?
a. *Agrobacterium*
b. *Rhizobium*
c. *Nitosomonas*
d. *Bacteroides*

31. Which of the following is true of *Agrobacterium tumefaciens*?
a. It is not capable of nitrogen fixation.
b. It causes crown gall disease when it carries a tumor inducing (Ti) plasmid.
c. It has been used to introduce foreign DNA into plant cells.
d. All of the above are true of *Agrobacterium tumefaciens*.

32. Members of the genus *Neisseria* cause which of the following human diseases?
a. gonorrhea
b. syphilis
c. genital herpes
d. vaginitis

33. Which of the following two genera is(are) nonmotile?
a. *Methylococcus*
b. *Methylomonas*
c. *Methylococcus* and *Methylomonas*
d. neither *Methylococcus* nor *Methylomonas*

34. Which of the following genera fix nitrogen nonsymbiotically?
a. *Rhizobium*
b. *Nitrosomonas*
c. *Nitrobacter*
d. *Azotobacter*

35. The similarity in appearance among the *Enterobacteriales* requires the use of biochemical ests for identification.

36. Members of the genus *Salmonella* cause
a. gastroenteritis.
b. typhoid fever.
c. gastroenteritis and typhoid fever.
d. neither gastroenteritis nor typhoid fever.

37. Which of the following organisms causes bubonic plague?
a. *Bordetella pertussis*
b. *Yersinia pestis*
c. *Helicobacter pylori*
d. *Campylobacter jejuni*

38. *Begiatoa* are
a. obligate aerobes.
b. facultative anaerobes.
c. microaerophiles.
d. obligate anaerobes.

39.TF The genus *Pseudomonas* contains more than 70 species.

40. The causative agent for cholera in humans is a member of the genus
a. *Escherichia*
b. *Vibrio*
c. *Enterobacter*
d. *Shigella*

41. *Escherichia coli* has not been found to be pathogenic in humans.

42. *Burkholderia cepacia* is a particular problem with
a. pneumonia patients.
b. AIDS patients.
c. myasthenia gravis patients.
d. cystic fibrosis patients.

43.TF *Rickettsia* have no cell walls.

44.TF Although myxobacteria are most abundant in warm areas, they will grow in arctic tundra.

45.TF Although discussed with the eubacteria, *Rickettsia* are obligate intracellular parasites and are therefore, more properly classified as viruses.

46. Members of the genus *Neisseria* cause which of the following diseases in humans?
a. gonorrhea
b. meningitis
c. gonorrhea and meningitis
d. neither gonorrhea nor meningitis

Chapter 23 Sample Test Questions on the Bacteria: The Low G + C Gram Positives

1. The enzyme _________ catalyzes the breakdown of hydrogen peroxide to water and oxygen.

2. The enzyme _________ causes blood plasma to clot.

3. Which of the following two genera is(are) photosynthetic using bacteriochlorophyll g?
   a. *Heliobacterium*
   b. *Heliophilum*
   c. Both a and b are correct.
   d. Neither a nor b is correct.

4. The gram-positive bacteria are important
   a. medically.
   b. ecologically.
   c. industrially.
   d. All of these.

5. The photosynthetic pigments of *Heliobacterium* and *Heliophilum*
   a. are in separate membrane structures attached to the plasma membrane.
   b. are in separate membrane structures that are not attached to the plasma membrane.
   c. are in the plasma membrane and not in separate structures.
   d. are in separate structures and also in the plasma membrane.

6. When peptidoglycan structure is analyzed for differences among various genera, the most significant difference(s) is (are)
   a. amino acid composition.
   b. linkage of the amino acids.
   c. both amino acid composition and linkage of the amino acids.
   d. neither amino acid composition nor linkage of the amino acids.

7. Members of the genus *Veillonella* are part of the normal microflora of the
   a. mouth.
   b. gastrointestinal tract.
   c. urogenital tract.
   d. mouth, gastrointestinal tract, and urogenital tract.

8. The Micrococcaceae divide in more than one plane to form
   a. regular clusters.
   b. irregular clusters.
   c. either regular or irregular clusters.
   d. neither regular nor irregular clusters. (They do not divide in more than one plane.)

9. *Thermoactinomyces vulgaris* is the causative agent for
    a. cystic fibrosis
    b. farmer's lung disease
    c. primary atypical pneumonia
    d. vaginitis

10. Because of its tolerance for high concentrations of sugar, *Leuconostoc* can cause spoilage of fruits packed in heavy syrup.

11. Mycoplasmas lack cell walls and cannot synthesize peptidoglycan precursors; therefore they
    a. are resistant to penicillin.
    b. are susceptible to lysis by osmotic shock.
    c. Both a and b are correct.
    d. Neither a nor b is correct.

12. Which of the following is(are) true of mycoplasmas?
    a. They are difficult to detect in mammalian cell cultures.
    b. They are a frequent contaminant of mammalian cell cultures.
    c. They are difficult to eliminate from mammalian cell cultures
    d. All of the above are true of mycoplasmas.

13. Primary atypical pneumonia in humans is caused by
    a. *Streptococcus pneumoniae*
    b. *Klebsiella pneumoniae*
    c. *Mycoplasma pneumoniae*
    d. *Pneumococcus pneumoniae*

14. Members of the genus *Staphylococcus* are normally associated with
    a. skin.
    b. skin glands.
    c. mucous membranes.
    d. All of these.

15. Abscesses and boils are usually associated with
    a. *Staphylococcus epidermidis.*
    b. *S. aureus.*
    c. either *Staphylococcus Epidermidis* or *S. aureus.*
    d. neither *Staphylococcus Epidermidis* nor *S. aureus.*

16. Toxic shock syndrome is produced by
    a. *Staphylococcus Epidermidis.*
    b. *S. aureus.*
    c. either *Staphylococcus Epidermidis* or *S. aureus.*
    d. neither *Staphylococcus Epidermidis* nor *S. aureus.*

17. Which of the following is a normal skin resident?
    a. *Staphylococcus epidermidis*
    b. *S. aureus*
    c. either *Staphylococcus epidermidis* or *S. aureus*
    d. neither *Staphylococcus epidermidis* nor *S. aureus*

18. The recently discovered giant bacterium is in the genus
    a. *Bacillus*
    b. *Clostridium*
    c. *Epulopiscium*
    d. *Desulfotomaculum*

19. Which of the following is responsible for food spoilage ,even in canned foods?
    a. *Clostridium perfringens*
    b. *Clostridium botulinum*
    c. *Clostridium acetobutylicum*
    d. *Clostridium tetani*

20. Members of the genus *Streptococcus* carry out __________ fermentation.
    a. butanediol
    b. heterolactic
    c. homolactic
    d. mixed acid

21. Some species of *Streptococcus*, when grown on blood agar, produce an incomplete hemolysis with the formation of a greenish zone around the colony; this is referred to as
    a. alpha-hemolysis.
    b. beta-hemolysis.
    c. gamma-hemolysis.
    d. None of these.

22. Some species of *Streptococcus*, when grown on blood agar, produce a complete hemolysis with the formation of a clear zone around the colony; this is referred to as
    a. alpha-hemolysis.
    b. beta-hemolysis.
    c. gamma-hemolysis.
    d. None of these.

23. Which of the following is associated with dental caries?
    a. *Streptococcus pneumoniae*
    b. *S. mutans*
    c. *S. fecaelis*
    d. *Lactococcus lactis*

24. Which of the following will grow in 6.5% salt?
    a. *Streptococcus mutans*
    b. *S. fecaelis*
    c. *Lactococcus lactis*
    d. All of these.

25. Which of the following is used in the production of buttermilk and cheese?
    a. *Streptococcus pneumoniae*
    b. *S. mutans*
    c. *S. fecaelis*
    d. *Lactococcus lactis*

26. Endospore-forming gram-positive rods and cocci are found primarily in __________ habitats.
    a. soil
    b. freshwater
    c. marine
    d. All of these.

27. Which of the following is currently being used or investigated for use as a biological insecticide?
    a. *Bacillus cereus*
    b. *B. subtilis*
    c. *B. anthracis*
    d. *B. thuringensis*

28. Which of the following is not normally considered a pathogenic species of *Clostridium*?
    a. *C. acetobutylicum*
    b. *C. botulinum*
    c. *C. perfringens*
    d. *C. tetani*

29. *Clostridium perfringens* causes which of the following?
    a. gas gangrene
    b. food poisoning
    c. both gas gangrene and food poisoning
    d. neither gas gangrene nor food poisoning

30. Members of the genus *Lactobacillus* are
    a. obligate aerobes.
    b. obligate anaerobes.
    c. either obligate aerobes or obligate anaerobes.
    d. neither obligate aerobes nor obligate anaerobes.

31. Which of the following is not true of the genus *Lactobacillus*?
    a. They are used in the production of fermented vegetable foods, beverages, sour dough, hard cheeses, yogurt, and sausages.
    b. They are responsible for spoilage of beer, milk, and meat.
    c. Many species are pathogenic in humans.
    d. All of these are true about the genus *Lactobacillus*.

32.TF Recently, viable endospores have been found that may be over 25 million years old.

33. *Thermoactinomyces* have very heat-resistant endospores; they can survive 90°C for
    a. 30 seconds
    b. 3 minutes
    c. 30 minutes
    d. 3 hours

34. Members of the genus *Lactobacillus* are normal microflora of the
    a. mouth.
    b. intestinal tract.
    c. vagina.
    d. All of the above are correct.

35.TF Members of the genus *Streptococcus* produce lactic acid but no gas.

36. Which of the following is not one of the groups of streptococci?
    a. the pyogenic streptococci
    b. the oral streptococci
    c. the anal streptococci
    d. the anaerobic streptococci

37. *Enterococcus faecalis* has been associated with
    a. endocarditis.
    b. urinary tract infections.
    c. endocarditis and urinary tract infections.
    d. neither endocarditis nor urinary tract infections.

38. Streptococcal sore throat (strep throat) is caused by
    a. *Streptococcus pyogenes*
    b. *Streptococcus mutans*
    c. *Streptococcus throatus*
    d. *Streptococcus pneumoniae*

40. *Clostridium tetani*, the causative agent of tetanus is a serious problem with
    a. puncture wounds.
    b. surface lacerations.
    c. puncture wounds and surface lacerations.
    d. neither puncture wounds nor surface lacerations.

41. *Listeria monocytogenes* causes listeriosis, an important
    a. waterborne infection.
    b. insect-borne infection.
    c. airborne infection.
    d. foodborne infection.

42. The largest genus in the order *Lactobacilliales* is ______________ with nearly 80 species.
    a. *Lactobacillus*
    b. *Leuconostoc*
    c. *Streptococcus*
    d. *Listeria*

43. Endocarditis is associated with
    a. *Staphylococcus aureus.*
    b. *Staphylococcus epidermidis.*
    c. both *Staphylococcus aureus* and *Staphylococcus epidermidis.*
    d. neither *Staphylococcus aureus* nor *Staphylococcus epidermidis.*

44. Which of the following is pyogenic (pus forming)?
    a. *Streptococcus pneumoniae*
    b. *Streptococcus mutans*
    c. *Streptococcus fecaelis*
    d. *Lactococcus lactis*

45. Members of the genus *Leuconostoc* are important in the production of
    a. wine.
    b. sauerkraut.
    c. buttermilk.
    d. All of these.

46. Which of the following genera produce useful antibiotics?
    a. *Bacillus*
    b. *Clostridium*
    c. *Leuconostoc*
    d. All of these.

47.TF Specific groups of gram-positive bacteria can sometimes be distinguished by differences in peptidoglycan structure.

48.TF Members of the genus Leuconostoc tolerate high sugar concentrations and will, therefore, grow in heavy syrup.

49. *Lactobacillus* grows optimally at
    a. slightly acidic pH.
    b. neutral pH.
    c. slightly alkaline pH.
    d. very acidic pH.

50. Which of the following antibiotics are produced by members of the genus *Bacillus*?
    a. bacitracin
    b. gramicidin
    c. polymyxin
    d. All of the above are produced by members of the genus *Bacillus*.

51.TF Lactobacilli are alkalophilic, preferring conditions that are slightly alkaline for optimal growth.

52.TF *Staphylococcus aureus* is coagulase positive which means it can cause blood plasma to clot.

53.TF Peptidoglycan structure differs among the gram-positive bacteria particularly in their amino acid content.

54.TF *Streptococcus fecaelis* has recently been reclassified into a separate genus called *Enterococcus*.

55.TF Members of the genus *Bacillus* are spore-forming, aerobic and catalase positive.

56.TF Members of the genus *Clostridium* are spore-forming, aerobic and catalase negative.

57. *Desulfotomaculum* are gram-positive but stain gram-negative because their cell walls have a lower than normal concentration of peptidoglycan.

58. Sporeforming bacteria are found mainly in
    a. freshwater environments.
    b. marine environments.
    c. soil environments.
    d. urban environments.

59. Members of the genus *Veillonella* are parasites that primarily inhabit
    a. cold-blooded animals.
    b. warm-blooded animals.
    c. warm-blooded and cold-blooded animals.
    d. neither warm-blooded nor cold-blooded animals; they are actually plant pathogens.

60.TF Many bacteria that were originally classified in the genus *Streptococcus* are now placed into two new genera, *Enterococcus* and *Lactococcus*.

Chapter 24 Sample Test Questions on the Bacteria: The High G + C Gram Positives

1. Actinomycetes are
    a. gram positive.
    b. gram negative.
    c. some are gram positive and others are gram negative.
    d. neither gram positive nor gram negative.

2.	Actinomycetes are
	a.	cocci.
	b.	bacilli.
	c.	curved.
	d.	filamentous.

3.	Which of the following is true of the actinomycetes?
	a.	Like other filamentous bacteria, the hyphae frequently undergo fragmentation.
	b.	Unlike other filamentous bacteria, the hyphae frequently undergo fragmentation.
	c.	Like other filamentous bacteria, the hyphae do not undergo fragmentation.
	c.	Unlike other filamentous bacteria, the hyphae do not undergo fragmentation.

4.	Actinomycetes form
	a.	substrate mycelia.
	b.	aerial mycelia.
	c.	both substrate mycelia and aerial mycelia.
	d.	neither substrate mycelia nor aerial mycelia.

5.	Some actinomycetes form a tissuelike mass called a
	a.	hypha.
	b.	sporangium.
	c.	thallus.
	d.	conidium.

6.	Actinomycetes form spores that are
	a.	asexual.
	b.	sexual, that is, they are used for sexual reproduction.
	c.	both asexual and sexual.
	d.	neither asexual nor sexual. (They do not form spores.)

7.	Which of the following is(are) true about actinomycete spores?
	a. They are heat resistant.
	b. They withstand desiccation well.
	c. They are heat resistant and withstand desiccation well.
	d. They are neither heat resistant nor do they withstand desiccation well.

8.	Actinomycete spores are usually formed on __________ mycelia.
	a.	aerial
	b.	substrate
	c.	both aerial and substrate
	d.	neither aerial nor substrate (They are formed in the absence of mycelia.)

9.	Actinomycete spores formed on the end of a filament, that is, not in a special structure, are called
	a.	endospores.
	b.	conidiospores.
	c.	sporangiospores.
	d.	myxospores.

10. Actinomycete spores formed within a special structure are called
    a. endospores.
    b. conidiospores.
    c. sporangiospores.
    d. myxospores.

11. The special structure in which actinomycetes form their spores is called a
    a. sporangium.
    b. conidium.
    c. sporatorium.
    d. basidium.

12. Which of the following is (are) not useful in classifying actinomycetes?
    a. mycelial morphology
    b. sporangia morphology
    c. color of the sporangia
    d. All of these are useful in classifying actinomycetes.

13. Which of the following cannot be used to help classify actinomycetes?
    a. cell wall type
    b. percent G + C content
    c. phospholipid composition of the membrane
    d All of these can be used to classify actinomycetes.

14. Actinomycetes are usually found in ___________ habitats.
    a. soil
    b. freshwater
    c. marine
    d. All of these.

15. Members of the genus *Mycobacterium* are said to be ___________ because basic fuchsin dye cannot be removed even with acid-alcohol treatment.

16. The cell walls of *Mycobacterium* contain waxes with 60-90 carbon ___________ acids.

17. Members of the genus *Micrococcus* are frequently pigmented with which of the following colors?
    a. yellow
    b. red
    c. orange
    d. All of the above are correct.

18. The actinoplanetes are usually found in _______ habitats.
    a. soil
    b. freshwater
    c. marine
    d. All of these.

19.   Actinoplanetes have extensive ___________ mycelia.
   a.   substrate
   b.   aerial
   c.   substrate and aerial
   d.   substrate nor aerial (They do not have mycelia.)

20.   The genera of actinoplanetes vary in
   a.   the arrangement of spores.
   b.   the development of spores.
   c.   both the arrangement and the development of spores.
   d.   neither the arrangement nor the development of spores.

21.   Which of the following antibiotics is (are) produced by actinoplanetes?
   a.   streptomycin
   b.   gentamicin
   c.   erythromycin
   d.   planomycin

22.   The streptomycetes represent ___________ of the viable organisms in the soil.
   a.   5%
   b.   10%
   c.   1-20%
   d.   1-10%

23.   The streptomycetes form conidiospores on their ___________ mycelia.
   a.   substrate
   b.   aerial
   c.   substrate and aerial
   d.   substrate nor aerial

24.   The streptomycetes form conidiospores by division of the aerial mycelia in
   a.   a single plane.
   b.   two planes simultaneously.
   c.   two planes consecutively.
   d.   multiple planes.

25.   The conidiospores of streptomycetes are
   a.   motile by means of flagella.
   b.   motile by means of cilia.
   c.   motile by means of pseudopods.
   d.   nonmotile.

26.   Maduromycetes produce ___________ of spores.
   a.   pairs
   b.   short chains
   c.   either pairs or short chains
   d.   neither pairs nor short chains

27. Members of the genus *Thermomonospora* produce __________ spores on their mycelia.
    a. single
    b. pairs of
    c. chains of
    d. clusters of

28. Members of the genus *Thermomonospora* produce spores on __________ mycelia.
    a. aerial
    b. substrate
    c. both aerial and substrate
    d. neither aerial nor substrate

29. Which of the following is correct about the genus *Arthrobacter*?
    a. They are rod-shaped during exponential growth.
    b. They are coccoid in stationary phase.
    c. Both a and b are correct.
    d. Neither a nor b is correct.

30. Diphtheria is caused by a member of which of the following genera?
    a. *Mycobacterium*
    b. *Corynebacterium*
    c. *Propionibacterium*
    d. *Nocardia*

31. Members of the genus *Mycobacterium* cause which of the following diseases in humans?
    a. tuberculosis
    b. leprosy
    c. tuberculosis and leprosy
    d. neither tuberculosis nor leprosy

32. *Nocardia* are found in which of the following habitats?
    a. soil
    b. aquatic
    c. both soil and aquatic
    d. neither soil nor aquatic

33. Which of the following is correct about members of the genus *Propionibacterium*?
    a. They contribute to the production of Swiss cheese.
    b. They cause acne vulgaris and contribute to the development of body odor.
    c. Both a and b are correct.
    d. Neither a nor b is correct.

34.TF Of the nearly 500 species in the genus *Streptomyces*, only *S. somaliensis* is known to be pathogenic in humans.

35.TF Septa divide the actinomycete mycelia into cells, each containing a single nucleus.

36.TF The sexual spores of actinomycetes are called sporangiospores and the asexual spores are called conidiospores.

37.TF Although the actinomycetes are generally nonmotile, the spores may be flagellated and, therefore, motile.

38.TF Although actinomycetes vary in many ways, they all share a common type of cell wall.

39.TF One way of distinguishing characteristics of actinomycetes that can be used to classify these organisms is the types of sugars found in extracts of the organisms.

40.TF Actinomycetes degrade a variety of organic compounds and are, therefore, important in mineralization processes.

41.TF A few species of actinomycetes have been found to be pathogenic in humans, other animals, and plants.

42.TF   Mycoplasmas are the smallest bacteria capable of self-reproduction.

43.TF Actinoplanetes that dwell in the soil play an important role in plant and animal decomposition.

44.TF   The maduromycetes differ from other high G + C gram-positives in that they contain the sugar madurose in cell extracts.

45. Members of the genus *Micrococcus* are usually
a. aerobic
b. anaerobic
c. facultatively anaerobic
d. microaerophilic

46. Members of the genus *Micrococcus* usually exist as
a. pairs
b. tetrads
c. irregular clusters
d. All of the above are correct.

47.TF Actinomycetes are of importance because they mineralize organic materials.

48.TF Actinomycetes produce many of the antibiotics we use in medicine.

49.TF Septa divide the actinomycete mycelia into long cells each containing several nucleoids.

50. *Nocardia* cause biodeterioration of rubber joints in water and sewer pipes.

51.TF Although the spores of actinomycetes are not completely heat resistant, they can be used to classify these organisms.

52.TF   The actinomycetes are capable of degrading a wide variety of organic acids.

53.TF   Most actinoplanetes are found in marine habitats.

54.TF   None of the nocardioform actinomycetes are pathogenic to humans.

55.TF   *Bifidobacterium bifidus* is a major pathogen in breast-fed babies.

56.   A(n) ______________ mycelium is a branching network of fibers beneath the surface of the agar growth medium.

57.   A(n) ______________ mycelium is a branching network of filaments that rises above the agar growth medium.

58.   Members of the genus *Micrococcus* are generally not pathogenic.

Answers to Sample Test Questions from Video 7

1. True
2. True
3. True
4. True
5. False
6. False
6. False
7. Cloning
8. True
9. True

Answers to Sample Test Questions from Text Chapter 21 (Bacteria: Deinococcci and Nonproteobacteria Gram Negatives) Web Site

1. Radiation and desiccaation
2. All of these groups of photosynthetic bacteria
3. Elementary
4. Both of these answers are correct
5. A gram-negative cell wall
6. Hyperthermophiles, *Thermogota,* and *Aquifex*
7. True
8. 30%
9. True
10. True
11. Lichen
12. True
13. True
14. Neither syphilis nor Lyme disease
15. Deeper, anaerobic zones of aquatic habitats

Answers to Sample Test Questions from Text Chapter 22 (Bacteria: The Proteobacteria) Web Site

1. All of these
2. True
3. True
4. *Rhizobium*
5. The breakdown of organic materials to inorganic materials.
6. A and B
7. True
8. True
9. Rocky Mountain spotted fever
10. Peptic ulcer disease
11. *Bordetella pertussis*
12. The space between the plasma membrane and cell wall.
13. 25%
14. *Vibrio* and *Proteobacter*
15. True
16. False
17. True
18. All of the above are true
19. *Azotobacter*
20. *Vibrio*

Answers to Sample Test Questions from Text Chapter 23 (Bacteria: The Lower G + C Gram Positives) Web Site

1. Clostridia, bacilli, and lactobacilli
2. True
3. True
4. True
5. Skin and mucous membranes
6. Farmer's lung disease
7. True
8. *Staphylococus epidermidis*
9. *Clostridium botulinum*
10. *Streptococcus faecalis*
11. Only a and b are true
12. True
13. Mouth
14. True
15. All of these
16. Soil environments

Answers to Sample Test Questions from Text Chapter 24 (Bacteria: The High G + C Gram Positives) Web Site

1. True
2. Soil
3. Both substrate mycelia and aerial mycelia
4. True
5. Asexual
6. True
7. 1-20%
8. True
9. *Corynebacterium*
10. True
11. True

Answers to Chapter 21 Sample Test Questions on the Bacteria: Deinococcci and Nonproteobacteria Gram Negatives

The questions in this test bank have been ranked for difficulty. The ratings run from 1 (for the easiest) to 8 (for the most difficult).

#	Answer/Diff	#	Answer/Diff
1	D / 3	28	T / 7
2	C / 6	29	F / 6
3	D / 7	30	T / 5
4	C / 5	31	T / 5
5	A / 7	32	B / 7
6	B / 7	33	C / 5
7	C / 5	34	C / 6
8	C / 4	35	C / 6
9	A / 6	36	C / 7
10	B / 6	37	C / 6
11	C / 8	38	T / 5
12	A / 4	39	T / 5
13	A / 7	40	T / 5
14	C / 4	41	T / 3
15	C / 9	42	T / 7
16	C / 8	43	D / 4
17	D / 7	44	C / 4
18	A / 6	45	C / 3
19	C / 7	46	T / 5
20	D / 6	47	A / 7
21	B / 5	48	T / 6
22	B / 6	49	D / 3
23	C / 5	50	phycoerythrin / 8
24	A / 5	51	C / 7
25	B / 8	52	C / 5
26	C / 8	53	C / 5
27	T / 6	54	phycobilisomes / 8
		55	F / 7

Answers to Chapter 22 Sample Test Questions on Bacteria: The Proteobacteria

The questions in this test bank have been ranked for difficulty. The ratings run from 1 (for the easiest) to 8 (for the most difficult).

#	Answer/Diff		#	Answer/Diff
1	A / 5		25	A / 6
2	B / 5		26	C / 5
3	sheath / 6		27	C / 7
4	B / 7		28	T / 4
5	C / 5		29	T / 8
6	B / 5		30	B / 4
7	D / 6		31	D / 5
8	colonial / 8		32	A / 3
9	C / 4		33	A / 4
10	swarmer / 6		34	D/ 7
11	D / 612		35	T / 4
12	A / 4		36	C / 4
13	B / 4		37	B / 5
14	C / 6		38	C / 7
15	B / 6		39	T / 5
16	C / 4		40	B / 6
17	A / 7		41	F / 3
18	B / 6		42	D/ 7
19	B / 5		43	F / 4
20	C / 5		44	T/ 6
21	T / 2		45	F / 7
22	D/ 6		46	C / 4
23	F / 3			

Answers to Chapter 23 Sample Test Questions on Bacteria: The Lower G + C Gram Positives

The questions in this test bank have been ranked for difficulty. The ratings run from 1 (for the easiest) to 8 (for the most difficult).

#	Answer/Diff	#	Answer/Diff
1	catalase / 6	31	C / 7
2	coagulase / 5	32	T / 7
3	C / 7	33	C / 6
4	D / 1	34	D / 8
5	C / 7	35	T / 4
6	C / 5	36	C / 2
7	D / 6	37	C / 4
8	C / 4	38	A / 5
9	B / 6	39	D / 7
10	T / 5	40	A / 4
11	C / 4	41	D / 6
12	D / 7	42	A / 2
13	C / 8	43	B / 7
14	D / 8	44	A / 5
15	B / 8	45	D / 5
16	B / 8	46	A / 5
17	A / 5	47	T / 5
18	C / 6	48	T / 5
19	B / 4	49	A / 5
20	C / 7	50	D / 6
21	A / 4	51	F / 3
22	B / 4	52	T / 5
23	B / 6	53	T / 5
24	B / 6	54	T / 7
25	D / 6	55	T / 8
26	A / 5	56	F / 8
27	D / 7	57	T / 7
28	A / 4	58	C / 3
29	C / 6	59	B / 5
30	D / 5	60	T / 6

Answers to Chapter 24 Sample Test Questions on Bacteria: The High G + C Gram Positives

The questions in this test bank have been ranked for difficulty. The ratings run from 1 (for the easiest) to 8 (for the most difficult).

#	Answer/Diff		#	Answer/Diff
1	A / 3		29	C / 6
2	D / 4		30	B /4
3	D / 7		31	C / 5
4	C / 5		32	C / 5
5	C / 5		33	C / 6
6	A / 6		34	T / 7
7	B / 7		35	F / 4
8	A / 5		36	F / 4
9	B / 6		37	T / 4
10	C / 6		38	F / 6
11	A / 5		39	T / 8
12	D / 7		40	T / 6
13	D / 7		41	T / 6
14	A / 4		42	T / 5
15	acid-fast / 5		43	T / 5
16	mycolic / 7		44	T / 3
17	D / 5		45	A / 4
18	D / 5		46	D / 5
19	A / 6		47	T / 4
20	C / 6		48	T / 3
21	B / 8		49	T / 5
22	C / 7		50	T / 5
23	B / 5		51	T / 8
24	A / 6		52	T / 5
25	D / 4		53	F / 7
26	C / 6		54	C / 5
27	A / 6		55	F / 7
28	C / 7		56	substarte / 3
			57	ariel / 3
			58	T / 4

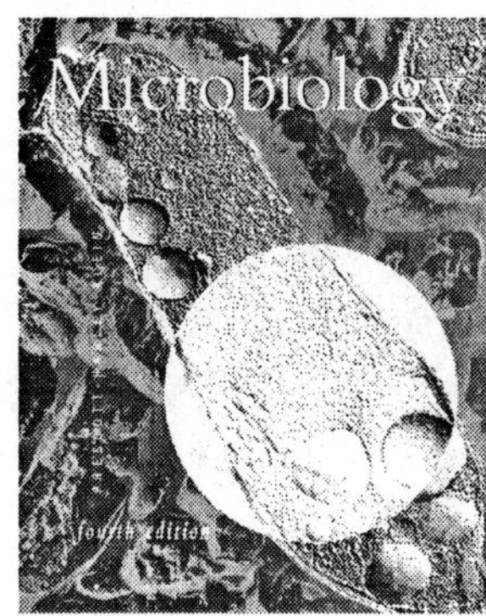

Microorganisms in the Environment
Program 8
Microbial Ecology

> Everything is everywhere, the environment selects.
> –M. W. Beijerinck

One of microbiology's fifty most significant events in the past 125 years.

1889

In the above year, **Martinus Beijerinck** used an enrichment culture, minus nitrogenous compounds, to obtain a pure culture of the root nodule bacterium *Rhizobium*, demonstrating that the enrichment culture creates the conditions for differential growth of a desired bacterium.

Student Assignments

- **Watch**   Video 8: Microbial Ecology
- **Read**   Chapter 40: Microorganisms as Components of Ecosystems
  Chapter 41: Marine and Freshwater Environments in *Microbiology*, 4th ed., Prescott/Harley/Klein, McGraw-Hill.
- **Lab Experiments** from *Laboratory Exercises in Microbiology*, 4th ed., Harley/Prescott, McGraw-Hill.
  Exercise 51: Standard Coliform Most Probable Number (MPN) Test and Presence-Absence Coliform Test
  Exercise 52: Membrane Filter technique for Coliforms and Fecal Streptococci: KONFIRM Test for Fecal Coliforms
  Exercise 75: Enumeration of Soil Microorganissms
  Exercise 76: Nitrogen Cycle

Lesson Overview

Video 8: *Microbial Ecology*
Humans and all other life forms depend on microorganisms as the essential processors of oxygen,

mineral nutrients for plant growth, and waste materials. Here we investigate some of the important environments dominated by microbes and how their presence is essential for human life.

Chapter 40: Microorganisms as Components of Ecosystems
This chapter discusses the roles of microorganisms in ecosystems. The organisms' physiological state, nutrient cycling, decomposition processes, and successional interactions are discussed. The fate and impact of genetically engineered organisms are also considered. The chapter includes a discussion of extreme environments and concludes with a discussion of the methods used in environmental studies.

Chapter 41: Marine and Freshwater Environments
This chapter discusses the general characteristics of microorganisms that are associated with marine and freshwater environments. Aquatic environments are dominated by the liquid phase, and the only oxygen available is that which can be dissolved in water. This dissolved oxygen can be quickly depleted, creating anaerobic conditions. The use of surface and subsurface waters for drinking water sources and for waste disposal is also discussed. The chapter concludes with a discussion of the contamination of ground waters by domestic and industrial wastes and the use of microorganisms to restore this important resource.

Learning Objectives

After successfully completing this unit, the student will be able to:

- discuss the populations and communities of microorganisms that are associated with various ecosystems environments and describe the roles they play in succession
- discuss the interactions of microorganisms with one another and with other nonmicrobial members of various ecosystems
- describe the ways in which microorganisms obtain their required nutrients from environmental sources
- discuss the fate and potential impact of the releasing genetically engineered microorganisms
- describe the microorganisms found in extreme environments and the adaptations necessary for their survival under these conditions
- discuss the various methods used to study microorganisms in the environment
- discuss the nature of marine and freshwater environments
- describe the complex microbial communities found in these environments
- discuss hydrothermal vents and hydrocarbon seeps
- discuss biofilms and the consequences of their formation
- discuss microbial mats
- discuss the recent concern over *Pfeisteria piscicida*, a protozoan parasite of fish
- discuss the influence of microorganisms in the oceans on carbon, nitrogen, phosphorus and sulphur cycling
- describe the possible cyclic changes in dissolved oxygen content of aquatic environments because of the unbalanced nature of oxygen depletion and replenishment
- discuss sewage treatment systems and their logical design according to the predictable sequences of the biological utilization of organic wastes

- describe the use of indicator organisms to measure the microbiological quality of water and describe the desirable characteristics of such an organism
- discuss the recent concern over the protozoan, *Cyclospora,* which causes long-lasting diarrheal disease
- discuss the limitations of indicator organisms with regard to the potential presence of viruses and protozoa in water sources
- discuss the ways that pathogenic organisms may be introduced into water sources other than by contamination with human wastes
- discuss the increasing concern about contamination of groundwater sources

Course Integrator

Chapter 40
Microorganisms and the Structure of Ecosystems (p. 832)

Visual Resource Library	Ecological Role of Microbes Figure 40.2
Transparency	Ecological Role of Microbes Figure 40.2
Visual Resource Library	Chemical Gradients in Nature Figure 40.5
Microbes in Motion II	Environmental Microbiology /Biogeochemical Recycling/Introduction, Habitat
Video *Unseen Life on Earth*	Unit #8 Microbial Ecology and Ecosystems
Web Link	*Microbiology* Online Learning Center: http://www.mhhe.com/biosci/cellmicro/prescott/
Web Link	The Curious Microbe: http://commtechlab.msu.edu/sites/dlc-me/curious/cindex.html
Web Link	The Microbial Zoo: http://commtechlab.msu.edu/sites/dlc-me/zoo/

The Physiological State of Microorganisms (p. 834)

Nutrient Cycling Processes (p. 836)

Visual Resource Library	The Carbon Cycle in Nature Figure 40.7
Transparency	The Carbon Cycle in Nature Figure 40.7
Visual Resource Library	The Sulfur Cycle in Nature Figure 40.8
Transparency	The Sulfur Cycle in Nature Figure 40.8
Visual Resource Library	The Nitrogen Cycle in Nature Figure 40.9
Transparency	The Nitrogen Cycle in Nature Figure 40.9
Visual Resource Library	The Mercury Cycle Figure 40.12
Microbes in Motion II	Environmental Microbiology/ Biogeochemical Recycling/Carbon Cycle Biogeochemical Recycling/Sulfur Cycle Biogeochemical Recycling/ Nitrogen Cycle

Interactions between Microorganisms and Metals (p. 840)

Interactions in Substrate Use (p. 842)

| *Microbes in Motion II* | Environmental Microbiology /Diversity/Detection in the Environment |
| *Microbes in Motion II* | Microbial Metabolism and Growth /Microbial Growth/ Metabolic Environment |

Organic Substrate Use by Microorganisms (p. 843)

Disease-Causing Microorganisms- Survival and Fate (p. 844)

Genetically Engineered Microorganisms- Fate and Effect (p. 845)

Extreme Environments (p. 846)

| Web Link | Halophilic Microorganisms: http://pasteur.bio.geneseo.edu/ |
| Web Link | Eukaryotes in extreme environments: http://www.nhm.ac.uk/zoology/extreme.html |

Methods Used in Environmental Studies (p. 847)

Chapter 41
The Nature of Marine and Freshwater Environments (p. 853)

Visual Resource Library	Confocal Scanning Microscopy Box 41.1
Visual Resource Library	The Winogradsky Column Figure 41.1
Transparency	The Winogradsky Column Figure 41.1
Visual Resource Library	The Growth of Biofilms Figure 41.3
Transparency	The Growth of Biofilms Figure 41.3
Visual Resource Library	Life Cycle of Rhizophydium Figure 41.6a
Microbes in Motion II	Environmental Microbiology/ Aquatic Habitats/Freshwater Aquatic Habitats/Marine, Deep Sea
Microbes in Motion II	Environmental Microbiology/ Microbial Ecology/Biofilms
Microbial Ecology/Microbial Mats	
Web Link	*Microbiology* Online Learning Center: http://www.mhhe.com/biosci/cellmicro/prescott/

The Microbial Community in Marine and Freshwater Environments (p. 858)

Visual Resource Library	Chytrid-Algal Interactions Figure 41.6d
Visual Resource Library	Ingoldian Fungi in a Stream Figure 41.7b
Microbes in Motion II	Environmental Microbiology/ Aquatic Habitats/Freshwater Aquatic Habitats/Marine, Deep Sea
Web Link	The Microbial Zoo (see WaterWorld) http://commtechlab.msu.edu/sites/dlc-me/zoo/

Water and Disease Transmission (p. 874)

Web Link   CDC Travel Information: Food and Water Precautions
http://www.cdc.gov/travel/foodwatr.htm)

Web Link   The "Bad Bug Book" from the US FDA
(http://vm.cfsan.fda.gov/~mow/intro.html)

Groundwater Quality and Home Treatment Systems (p. 880)

Visual Resource Library	Treatment Plant Overview Figure 41.17b
Visual Resource Library	Secondary Sewage Treatment Figure 41.20
Visual Resource Library	A Constructed Wetland System Figure 41.23
Visual Resource Library	Water Purification Steps Figure 41.24
Transparency	Water Purification Processes Figure 41.24
Visual Resource Library	Multiple-Tube Water Testing Figure 41.26
Transparency	Multiple-Tube Water Testing Figure 41.26
Microbes in Motion II	Environmental Microbiology/

Harness for Production/Solid Waste Treatment
Harness for Production /Water Purification

Microbes in Motion II	Parasitology/Protozoa Classification /Sporozoan Protozoa
Microbes in Motion II	Parasitology/Intestinal Protozoa
Web Link	Bacteria and Other Microorganisms in Household Water http://www.cdc.gov/niosh/nasd/docs4/va98020.html
Hyperclinic	Body Systems/Gastrointestinal System

Microbial Terms Used in Video 8 in Order of Appearance with Page Indexing to the Textbook (*Microbiology*, 4th ed., Prescott/Harley/Klein, McGraw-Hill)

ecosystem 834	decomposer 524
photosynthesis 182	nitrogen fixation 198
food web 554	rumen 570
primary producer 832	biodegradation 952

Web Links Related to Video 8 (Microbial Ecology) and Readings in Chapter 40 (Microorganisms as Components of Ecosystems) and Chapter 41 (Marine and Freshwater Environments) (Go to www.mhhe.com/prescott, Student on Line Learning Center, Double Click on Chapter 40 [and then 41], Double Click on Web Sites)

Chapter 40

The Global Carbon Cycle
http://www.esd.ornl.gov/iab/iab2-2.htm

What is the Carbon Cycle
http://library.advanced.org/11226/why.htm

The Nitrogen Cycle
http://markw.com/ncycle.htm

Introducing Fish, Plants, and Food to Your Aquarium Begins a Natural Process Called the Nitrogen Cycle
http://www.algone.com/cycle.htm
The Nitrogen Cycle in the Aquarium
http://home.earthlink.net/~grenier2/nitro.htm

Rock-Eating Microbes
http://www.accessexcellence.org/WN/SU/rockeater898.html

Life in the Deep Freeze
http://www.accessexcellence.org/WN/SUA12/marscold698.html

From Primordial Soup to the Prebiotic Beach. An interview with exobiology pioneer, Dr. Stanley Miller, University California, San Diego.
http://www.accessexcellence.org/WN/NM/miller.html

Eukaryotes in extreme Environments
http://www.nhm.ac.uk/zoology/extreme.html#Anaerobes

Chapter 41

CDC Travel Information Food and Water Precautions
http://www.cdc.gov/travel/foodwatr.htm

Bacteria and Other Microorganisms in Household Water
http://www.cdc.gov/niosh/nasd/docs4/va98020.html

Microbial Terms Used in Chapter 40 (Microorganisms as Components of Ecosystems)

allochtonous 844	demand (BOD) 848	chemical oxygen
assimilatory reduction 837	biofilms 835	demand 848
	biogeochemical	community 834
bacteriostasis 835	cycling 836	consumer 832
biochemical oxygen	biomagnification 841	decomposer 832

denitrification 838
dissimilatory
  reduction 837
ecosystem 832
extreme barophilic
  bacteria 846
food web 833
fungistasis 835
gradient 834
heterotrophic
  nitrification 838

Liebig's law of the
  the minimum
microbial ecology
  834
microbiostasis 835
microenvironment 834
mineralization 833
moderately barophilic
  bacteria 846
nitrification 838
oliootrophic
environments 835

population 833
primary producer 832
primary production
  832
Shelford's law of
tolerance 834
sulfate reduction
  837
total organic carbons
  848

Microbial Terms Used in Chapter 41 (Marine and Freshwater Environments)

anaerobic digestion
  871
activated sludge 869
barophile 860
biofilms 855
bulking sludge 869
coagulation 874
constructed wetlands
  873
defined substrate test
  878
diurnal oxygen shifts
  867
epiliminion 866
eutrophic 866
eutrophication 866
extended aeration
870
fecal coliforms 878
fecal enterococci 878
groundwater 880
hypolimnion 866

indicator organisms
  876
Inglodian fungi 858
in situ treatment 852
membrane filtration
technique 878
microbial mat 856
most probable
number 878
MUG 879
nitrogen oxygen
demand 868
nonpoint source of
polution 867
oligotrophic 866
ONPG 879
phytoplankton 862
point source of
polution 867
potable 876

presence-absence test
  878
primary treatment
869
rapid sand filter 874
secondary treatment
  869
sedimentation basin
  874
septic tank 880
settling basin 874
sewage treatment 869
slow sand filter 874
sludge 869
tertiary treatment 873
thermocline 866
total organic carbon
  868
trickling filter 870
ultramicrobacteria
  862
Winogradsky column
  853

Sample Test Questions from Video 8 (Microbial Ecology)

1. Ecosphere I was created by NASA as a step toward building a self-sustainable atmosphere for future space travel.
   a. True
   b. False

2. The total community of living organisms and their associated chemical and physical environments is called an __________ .

3. A network of interlinked food chains, encomposing primary producers, consumers, decomposers, and detritivores is called a __________ .

4. Every food web begins with a
   a. consumer
   b. producer
   c. herbivore
   d. carnivore

5. Aquatic environments are not as productive, acre per acre, as are terrestrial environments.
   a. True
   b. False

6. A cows rumen operates under __________ conditions.
   a. aerobic
   b. anaerobic

7. Masses of __________ bacteria swarm on the root hairs of legumes.

8. Chemoheterotrophs and chemoautotrophs exists together in a single sludge system.
   a. True
   b. False

9. The use of biologically mediated processes to remove or degrade pollutants from specific environments is called __________ .

10. Methane eating bacteria are called __________ .

Sample Test Questions from Text Chapter 40 (Microorganisms as Components of Ecosystems) Web Site

1. Primary production can be best described as the process of
   O   organic matter formation
   O   organic matter decomposition
   O   inorganic matter decomposition

2. Chemoautotrophs can contribute to primary production because they can
   O   fix carbon dioxide in the light.
   O   fix carbon dioxide without light energy.
   O   decompose organic matter in the process of mineralization.

3. The law of limiting factors or law of the minimum is called
   - O   Smith's law
   - O   Jones' law
   - O   Liebig's law

4. Lignin is degraded only under anaerobic conditions.
   - O   True
   - O   False

5. The process of nitrification involves the transformation of
   - O   ammonia to organic matter nitrogen.
   - O   nitrogen gas to ammonia.
   - O   nitrate to nitrogen gas.
   - O   ammonia to nitrate.

6. The process of nitrogen fixation involves the transformation of
   - O   ammonia to organic matter nitrogen.
   - O   nitrogen gas to ammonia.
   - O   Nitrate to nitrogen gas.

7. In the process of biomagnification, higher relative concentrations of a compound is found in
   - O   higher consumers as opposed to primary producers.
   - O   primary producers as opposed to higher consumers.

8. Microorganisms that are normally present in a particular environment are called indigenous.
   - O   True
   - O   False

9. How do microorganisms respond to low nutrient levels and intense competition?
   - O   increase in cell surface area
   - O   alteration of morphology to form mini or ultramicro cells
   - O   increase in attachments to certain surfaces
   - O   all of these are responses of microorganisms to low nutrient levels and intense competition

10. Chemoautotrophs serve as primary producers in
   - O   terrestrial soils
   - O   shallow freshwater environments
   - O   deep marine seeps of hydrogen sulfide

11. Microorganisms primarily function as
   - O   primary producers
   - O   consumers
   - O   decomposers
   - O   pathogens

12. Extreme environments will usually have a more diverse microbial community than a less extreme environment.
    O   True
    O   False

13. Ecosystems can function effectively for long periods without decomposition processes occurring.
    O   True
    O   False

14. Microorganisms always find ample amounts of nutrients available for their growth in natural environments.
    O   True
    O   False

15. All living microorganisms found in soils and waters are capable of growth on common laboratory media.
    O   True
    O   False

16. Genetically engineered microorganisms grown in the laboratory are usually less capable of survival in natural environments than naturally occurring microorganisms.
    O   True
    O   False

17. The possibility of gene transfers from genetically engineered microorganisms to natural populations is because of
    O   the existence of naturally competent microorganisms
    O   the stabilization of DNA on solid surfaces
    O   both a and b

Sample Test Questions from Text Chapter 41 (Marine and Freshwater Environments) Web Site

1. Oxygen is present in waters in the general range of
    O   kilograms per liter
    O   grams per liter
    O   milligrams per liter

2. Sessile microorganisms are most usually found associated with surfaces.
    O   True
    O   False

3. How will the solubility of oxygen in water change when the water temperature is lowered?
   - O   no change
   - O   increase
   - O   decrease

4. When algae use dissolved carbon dioxide present in waters, the water pH will
   - O   increase
   - O   decrease
   - O   not be changed

5. In the biochemical oxygen demand (BOD) test, the factor that should limit biological activity is
   - O   nitrogen
   - O   phosphorus
   - O   oxygen
   - O   carbon

6. Which procedure(s) can be used to measure microbial biomass in water?
   - O   chlorophyl levels
   - O   dry weight of filtered biomass
   - O   ATP
   - O   all of these.

7. Eutrophication is the process of acidification of waters.
   - O   True
   - O   False

8. If phosphorus is added to a water body that contains no other major nutrients, the major microbial group that will respond will be the
   - O   green sulfur bacteria.
   - O   green nonsulfur bacteria.
   - O   green algae.
   - O   cyanobacteria.

9. The process in which diurnal oxygen shifts can occur in a river after organic matter is added is called self-purification.
   - O   True
   - O   False

10. Secondary treatment of sewage involves the use of which type of process?
   - O   biological
   - O   physical/chemical
   - O   both biological and physical/chemical

11. *Giardia*, the cause of an important waterborne disease, is a
    O   virus
    O   bacterium
    O   fungus
    O   protozoan

12. Fecal coliforms are useful indicator organisms for the contamination of water by
    O   enteric viruses
    O   enteric bacteria
    O   enteric protozoan pathogens
    O   all of these

13. Removal of 99% of the organisms in drinking water will be sufficient to avoid disease transmission.
    O   True
    O   False

Sample Test Questions from Text Chapter 40 (Microorganisms as Components of Ecosystems)

1. Primary production can be best described as the process of
    a. organic matter formation.
    b. organic matter decomposition.
    c. inorganic material formation.
    d. inorganic material decomposition.

2. A simple self-regulating ecosystem can be constructed with
    a. higher consumers alone.
    b. only decomposers that mineralize organic matter.
    c. decomposers and primary producers.
    d. None of these.

3. Chemoautotrophs can contribute to primary production because they can
    a. fix carbon dioxide in the light.
    b. fix carbon dioxide without light energy.
    c. decompose organic matter in the process of mineralization.
    d. None of these.

4. Limiting factors for microbial growth can include
    a. light and light quality.
    b. carbon and nutrients.
    c. temperature and pressure.
    d. All of these.

5. On a dry weight basis, microbial cells typically consist of
   a. 10% carbon, 3% nitrogen, no phosphorus.
   b. 50% carbon, 14% nitrogen, 3% phosphorus.
   c. only carbon.
   d. None of these.

6. The law of limiting factors or law of the minimum is called
   a. Smith's law.
   b. Jones' law.
   c. Liebig's law.
   d. None of these.

7. Biogeochemical cycling includes which of the following?
   a. biological processes
   b. chemical processes
   c. Both a and b are correct.
   d. Neither a nor b is correct.

8. Biogeochemical cycles that contain gaseous components include
   a. the carbon, nitrogen, and phosphorus cycles.
   b. the carbon, nitrogen, and sulfur cycles.
   c. the carbon, nitrogen, phosphorus, and iron cycles.
   d. None of these.

9. A major fully reduced form of carbon, important in nutrient cycling is
   a. carbon dioxide.
   b. organic matter.
   c. methane.
   d. hydrogen sulfide.

10. Utilization of sulfur and nitrogen can involve either assimilation or dissimilation processes. Which of the following is an assimilation process?
    a. nitrate reduction to gaseous nitrogen and nitrous oxide
    b. sulfate reduction to hydrogen sulfide
    c. nitrate reduction to cellular amino acid nitrogen
    d. All of these.

11. Which of the following is a nitrifying genus?
    a. *Thiobacillus*
    b. *Escherichia*
    c. *Nitrobacter*
    d. *Azotobacter*

12. The process of nitrification involves the transformation of
    a. ammonia to organic matter nitrogen.
    b. nitrogen gas to ammonia.
    c. nitrate to nitrogen gas.
    d. ammonia to nitrate.

13. The process of denitrification involves the transformation of
    a. ammonia to organic matter nitrogen.
    b. nitrogen gas to ammonia.
    c. nitrate to nitrogen gas.
    d. ammonia to nitrate.

14. The process of nitrogen fixation involves the transformation of
    a. ammonia to organic matter nitrogen.
    b. nitrogen gas to ammonia.
    c. nitrate to nitrogen gas.
    d. ammonia to nitrate.

15. Methylated metals have a greater ability to pass the blood-brain barrier. A metal that can form stable methylated forms is
    a. Hg (Mercury).
    b. Cu (copper).
    c. Fe (iron).
    d. Zn (zinc).

16. Biomagnification involves compounds having higher relative concentrations in
    a. higher consumers as opposed to primary producers.
    b. primary producers as opposed to higher consumers.
    c. neither higher consumers nor primary producers.
    d. None of these.

17. *Desulfovibrio* and *Methanobacterium* can compete for the hydrogen present in their environment. This competition is dependent on the available oxidants, including
    a. sulfate and carbon dioxide.
    b. nitrate and hydrogen sulfide.
    c. methane and nitrogen gas.
    d. None of these.

18. The major organic substrates that contain nitrogen include
    a. cellulose and protein.
    b. starch and lignin.
    c. hydrocarbons and lignin.
    d. proteins and chitin.

19. Peat bogs occur in flooded areas because the rate of decomposition of which compound will be slower under anaerobic conditions?
    a. starch
    b. lignin
    c. protein
    d. lipids

20. Microorganisms that are not normally present in a particular environment are called
    a. indigenous.
    b. autochthonous.
    c. allochthonous.
    d. endogenous.

21. Extreme environments will usually have a __________ diverse microbial community than a less extreme environment.
    a. more
    b. less
    c. similarly

22. The survival of intestinally derived bacteria in soils and waters will usually be __________ as temperatures are lowered.
    a. increased
    b. decreased
    c. not changed

23. Microorganisms that can grow on cellulose as a substrate in media that do not contain inorganic mineral nitrogen in solution are usually capable of
    a. nitrification.
    b. nitrogen dissimilation.
    c. denitrification.
    d. nitrogen fixation.

24. Under aerobic conditions, the major forms of sulfur that accumulate as a result of microbial activities will usually be
    a. oxidized.
    b. reduced.
    c. neither oxidized nor reduced.
    d. oxidized and reduced.

25. The oxidized form of inorganic iron that results from microbial oxidative processes is
    a. ferrous ion.
    b. ferric ion.
    c. sulfate ion.
    d. nitrate ion.

26. Sulfide oxidation can occur rapidly in the presence of oxygen without microorganisms being present, which will result in the accumulation of
    a. nitrogen gas.
    b. elemental sulfur.
    c. sulfate.
    d. nitrate.

27. Which of the following has been made possible because of the differing sensitivities of microorganisms and more complex organisms to metals and metalloids?
    a. biological processing of mining ores to liberate useful material
    b. many of the antiseptic procedures currently in use
    c. Both of these answers are correct.
    d. Neither of these answers is correct.

28. Which of the following contributes to microbial succession?
    a. waste products from one organism or group of organisms will stimulate the growth of other organisms that can utilize those waste products as nutrients
    b. mixtures of electron acceptors will be used in a defined sequence by different sets of microorganisms
    c. Both of these answers are correct.
    d. Neither of these answers is correct.

29. Which of the following is not a response of microorganisms to low nutrient levels and intense competition?
    a. an increase in cell surface area
    b. alteration of morphology to form mini or ultramicro cells
    c. an increase in attachments to certain surfaces
    d. All of these are responses of microorganisms to low nutrient levels and intense competition.

30. In which of the following environments do chemoautotrophs serve as primary producers?
    a. terrestrial soils
    b. shallow freshwater environments
    c. deep marine seeps of hydrogen sulfide
    d. None of these.

31. Although microorganisms can serve a variety of roles they primarily function as
    a. primary producers.
    b. consumers.
    c. decomposers.
    d. top level consumers.

32. The metabolic products of one group of microorganisms
    a. are always toxic to other microorganisms.
    b. are sometimes toxic to other microorganisms.
    c. are seldom toxic to other microorganisms.
    d. are never toxic to other microorganisms.

33. All microbial substrates can be degraded under
   a. aerobic conditions.
   b. anaerobic conditions.
   c. aerobic and anaerobic conditions.
   d. either aerobic or anaerobic conditions but not both.

34. Microorganisms
   a. are a good nutrient source.
   b. will be degraded by other microorganisms.
   c. are a good nutrient source and will be degraded by other microorganisms.
   d. are neither a good nutrient source nor will be degraded by other microorganisms.

35. Hydrogen and methane oxidation occurs
   a. more rapidly under aerobic conditions.
   b. more rapidly under anaerobic conditions.
   c. with equal rapidity under aerobic and anaerobic conditions.
   d. not at all.

36.TF Wood pilings that remain below the water table are generally not subject to biodegradation because the conditions there are anaerobic and the organisms that carry out lignin degradation are obligate aerobes.

37.TF Heterotrophic nitrification is carried out by bacteria but not by fungi.

38.TF Chemolithotrophic microorganisms are infrequently found in natural environments.

39.TF Most microorganisms in soils and waters are spore formers because formation of spores allows them to survive better than formation of vegetative cells in these low-nutrient environments.

40.TF Hydrocarbons can serve as an excellent source of phosphorus for microbial growth.

41.TF "Noble" metals, such as gold and silver, can be used to control microbial growth.

42.TF The terms microbial populations and microbial communities mean the same thing.

43.TF Microorganisms always compete and never help one another in growth and utilization of substrates.

44.TF *Desulfovibrio* plays a major role in sulfur oxidation processes in soils and water.

45.TF Denitrification is a dissimilatory process by which nitrate is utilized by microorganisms.

46.TF Ecosystems can function effectively for long periods without decomposition processes occurring.

47.TF Higher consumers, such as humans, can contribute to primary production.

48.TF Microorganisms always find ample amounts of nutrients available for their growth in natural environments.

49.TF Nitrogen fixation is an energy-yielding process.

50.TF All living microorganisms found in soils and waters are capable of growth on common laboratory media.

51.TF Genetically engineered microorganisms grown in the laboratory are usually less capable of survival in natural environments than naturally occurring microorganisms.

52.TF The internal pH of acidophiles and alkalophiles will be similar to the external pH for each of these microbial groups.

53.TF *Sulfolobus* is a well-known thermophilic microorganism found in hot springs.

54.TF Microbial growth has been observed only in environments in which water is present in the liquid phase below 100 degrees C.

55.TF Thermophiles can be expected to have many more enzymes that will be more heat-stable than those found in mesophiles.

56. Which of the following contributes to the possibility of gene transfers from genetically engineered microorganisms to natural populations?
   a.   the existence of naturally competent microorganisms
   b.   the stabilization of DNA on solid surfaces
   c.   Both a and b are correct.
   d.   Neither a nor b is correct.

57. Degradation of organic material by microorganisms is influenced by
   a.   the composition and structure of the organic material
   b.   the microbial community present
   c.   Both a and b are correct.
   d.   Neither a nor b is correct.

58.TF Most microorganisms in natural environments are confronted with nutrient deficiencies that limit their activities.

59. Which of the following is true about genosensors?
   a. They are gel array microchips containing a mixture of probes for detecting specific microorganisms in a mixed population.
   b. They are used to detect organisms in environmental samples that cannot be cultured in the laboratory.
   c. Both a and b are correct.
   d. Neither a nor b is correct.

60. Which of the following is(are) used by aerobic nitrogen-fixing bacteria to maintain the anaerobic conditions needed for the nitrogen reduction reactions?
a. physical barriers to keep O_2 out of certain compartments
b. O_2 - scavenging molecules to remove oxygen
c. high metabolic rates to utilize oxygen as quickly as it becomes available
d. All of the above are correct.

Sample Test Questions from Text Chapter 41 (Marine and Freshwater Environments)

1. Oxygen is present in waters in the general range.
a. kg/liter.
b. ng/liter.
c. grams/liter.
d. mg/liter.

2. Sessile, prosthecate, and gliding microorganisms are most usually found associated with
a. air bubbles.
b. surfaces.
c. neither air bubbles nor surfaces.
d. air bubbles and surfaces.

3. The solubility of oxygen in water will change in which way when the water temperature is lowered?
a. No change will occur.
b. The solubility will increase.
c. The solubility will decrease.
d. The solubility will first decrease then increase.

4. At acidic pH values, the dissolved carbon-dioxide-bicarbonate-carbonate equilibrium in water will yield predominantly
a. dissolved carbon dioxide.
b. bicarbonate.
c. carbonate.
d. equal amounts of each.

5. When microorganisms, such as algae, use dissolved carbon dioxide present in waters, the water pH will
a. increase.
b. decrease.
c. not be changed.
d. first increase then decrease.

6. In the biochemical oxygen demand (BOD) test, the factor that should limit biological activity is
   a. nitrogen.
   b. phosphorus.
   c. oxygen.
   d. carbon.

7. In the biochemical oxygen demand (BOD) test, bottles to be analyzed should have which of the following oxygen changes in mg/liter?
   a. 2 used, 1 remaining
   b. 1 used, 2 remaining
   c. 10 used, 1 remaining
   d. none used

8. Which of the following procedure(s) can be used to measure microbial biomass in water?
   a. chlorophyl levels
   b. dry weight of filtered biomass
   c. ATP
   d. All of these.

9. Eutrophication is the process of
   a. nutrient enrichment of waters.
   b. nutrient removal from waters.
   c. acidification of waters.
   d. deoxygenation of waters.

10. If phosphorus is added to water that contains no other major nutrients, the major microbial group that will respond if the water is exposed to light will be the
    a. green sulfur bacteria.
    b. green nonsulfur bacteria.
    c. green algae.
    d. cyanobacteria.

11. The process in which diurnal oxygen shifts can occur in a river after organic matter is added is called
    a. self-reduction.
    b. self-purification.
    c. self-oxidation.
    d. self-eutrophication.

12. Secondary treatment of sewage involves the use of which type of process?
    a. biological
    b. physical/chemical
    c. both biological and physical/chemical
    d. neither biological nor physical/chemical

13. Activated sludge is used in which sewage treatment step?
    a. primary
    b. secondary
    c. tertiary
    d. All of these.

14. Slow rate biological processing in biological sewage treatment will produce effluents that are
    __________ those produced by higher rate systems.
    a. better than
    b. worse than
    c. the same as

15. Filamentous organisms such as *Sphaerotilus* and *Thiothrix* can lower the efficiency of activated
    sludge processes by causing
    a. anaerobic sludge.
    b. aseptic sludge.
    c. bulking sludge.
    d. All of these.

16. A substrate(s) for production of methane by methanogens include(s)
    a. ethanol.
    b. hydrogen and carbon dioxide.
    c. glucose.
    d. citric acid.

17. Once formed in the sewage treatment biological process, the microbial biomass will degrade in
    extended aeration due to the process.
    a. endogenous respiration.
    b. anabolism.
    c. eutrophication.
    d. heterotrophic nitrification.

18. *Giardia* is an important waterborne disease. This organism is a
    a. virus.
    b. bacterium.
    c. fungus.
    d. protozoan.

19. Sewage sludge disposal can lead to accumulation in the soil of
    a. heavy metals.
    b. pesticide residues.
    c. neither heavy metals nor pesticide residues.
    d. heavy metals and pesticide residues.

20. Indicator organisms are used in analyses of drinking water because
    a. the indicators die off slower than the disease organisms.
    b. the indicators die off faster than the disease organisms.
    c. the indicators and the disease organisms both die off at the same rates.
    d. None of these.

21. The presumptive test involves detection of
    a. gas production in a lactose medium.
    b. sheen production on a selective medium.
    c. a gram-negative non spore-forming rod.
    d. All of these.

22. The completed test involves detecting
    a. gas production in a lactose medium.
    b. sheen production on a selective medium.
    c. the presence of a gram-negative nonspore-forming rod.
    d. None of these.

23. The fecal enterococci are often used as an indicator in which of the following?
    a. brackish and marine water
    b. lake and river water
    c. groundwater
    d. None of the above are correct.

24. The presence-absence test is a modification of which test?
    a. membrane filter
    b. MPN
    c. virus plaque assay
    d. None of these.

25. *Cryptosporidium* is a major cause of enterocolitis. The detection of this organism, normally present in water, is especially important in
    a. infants.
    b. immunologically compromised individuals.
    c. the aged.
    d. None of these.

26. Coliforms and fecal coliform are often differentiated based on their responses to
    a. temperature.
    b. salt tolerance.
    c. pressure.
    d. metal sensitivity.

27. A columnar set-up designed to resemble a naturally occurring water/sediment ecosystem is called a(n)
    a. microcosm.
    b. mesocosm.
    c. macrocosm.
    d. None of these.

28. In which of the following types of waters do seasonal climatic temperature changes result in distinct chemical and microbial stratification?
    a. oceans
    b. eutrophic lakes
    c. oligotrophic lakes
    d. fast flowing rivers

29. Sewage treatment that involves horizontal flow in an agitated aeration tank is referred to as
    a. lagooning.
    b. activated sludge treatment.
    c. trickling filter processing.
    d. endogenous respiration.

30. Which of the following is not an advantage of anaerobic sludge digestion?
    a. most of the biomass produced aerobically is utilized for methane production
    b. the remaining sludge can be dried easily before disposal
    c. heavy metals are concentrated in the sludge
    d. All of these are advantages of anaerobic sludge digestion.

31. Fecal coliforms are useful indicator organisms for the contamination of water by
    a. enteric viruses.
    b. enteric bacteria.
    c. enteric protozoan pathogens.
    d. All of these.

32. Which of the following does not contribute to the removal of organic material in home sewage treatment system leach fields?
    a. aerobic digestion as the waste percolates through the soil
    b. adsorption of organic material to soil particle surfaces
    c. entrapment of microbes in the pores of the leach field
    d. All of these contribute to the removal of organic material.

33. Which of the following is not true about microbial mats?
    a. they form on the surfaces of rocks and sediments
    b. anaerobic conditions prevail in parts of the mat where light does not penetrate
    c. gradients are formed that are very gradual with great distance between recognizable layers
    d. All of these are true about microbial mats.

34. Primary treatment of sewage involves the use of which type of process?
   a. biological
   b. physical/chemical
   c. both biological and physical/chemical
   d. neither biological nor physical/chemical

35.TF A septic tank will function effectively without a leach field.

36.TF Light cannot penetrate into water that does not contain oxygen.

37.TF Microorganisms in water can utilize oxygen faster than it can be redissolved if enough nutrients are available.

38.TF Low-nutrient, oligotrophic waters will usually become stratified and have an anaerobic hypolimnion at certain times of the year.

39.TF In the total organic carbon (TOC) test, microorganisms are used to measure the levels of carbon in a water sample.

40.TF The biochemical oxygen demand (BOD) test is primarily a measure of carbonaceous demand and should not include nitrogen oxidation demand.

41.TF The epilimnion of a lake is located below the hypolimnion.

42.TF In general, cyanobacteria prefer higher pH and temperature conditions than algae.

43.TF If organic acids accumulate in an anaerobic digester, a stuck digester can result.

44.TF Organisms, such as Methanobacterium, carrying out the final step in methane production are facultative anaerobes.

45.TF Limits on COD and BOD levels in treated sewage have been established in order to prevent eutrophication of outfall areas.

46.TF If a leach field floods it becomes anaerobic and effective treatment ceases.

47.TF Potable water is unfit for consumption or recreation because of the high levels of microbial contaminants present.

48.TF Removal of 99% of the organisms in drinking water will be sufficient to avoid disease transmission.

49.TF Rapid sand filtration, used in many water purification plants, is sufficient to ensure that *Giardia* will be consistently removed.

50.TF Complex organic polymers can be transformed directly into methane by methanogens.

51.TF Chlorination will usually provide sufficient oxidizing power to inactivate viruses present in drinking water.

52.TF Indicator organisms are added to drinking water to test whether the purification process is being carried out effectively.

53.TF All coliforms can grow at 44.5° C.

54.TF *Cryptosporidium* has recently become of greater concern than *Giardia* because it occurs more often in waters and is harder to remove.

55.TF The membrane filter technique can be used in all waters and has completely replaced the MPN technique.

56. Constructed wetlands use plants and their associated microorganisms to remove which of the following?
   a. organic material
   b. inorganic material
   c. metals
   d. All of the above.

57.TF Biofilms are becoming of increasing concern and may be a major cause of such pathogens as Legionella in potable water systems.

58. Which of the following occurs in large numbers in marine environments?
   a. barotolerant/barophilic bacteria
   b. viruses
   c. archaeobacteria
   d. All of the above.

59. Slow sand filtration effectively removes
   a. *Giardia lamblia.*
   b. *Cryptosporidium parvum.*
   c. both *Giardia lamblia* and *Cryptosporidium parvum.*
   d. neither *Giardia lamblia* nor *Cryptosporidium parvum.*

60.TF Despite their small size, microorganisms in the ocean can have a significant impact on the atmosphere on a global scale.

Answers to Sample Test Questions from Video 8

1. True
2. Ecosystem
3. Food web
4. B
5. False

6. B
7. rhizobial
8. True
9. Bioremediation
10. Methanotrophs

Answers to Sample Test Questions from Text Chapter 40 (Microorganisms as Components of Ecosystems)

1. Organic matter formation
2. Fix carbon dioxide without light energy
3. Liebig's law
4. False
5. Nitrate to nitrogen gas
6. Nitrogen gas to ammonia
7. Higher consumers as opposed to primary producers
8. True
9. All of these

10. Deep marine seeps of hydrogen sulfide
11. Decomposers
12 False
13. False
14. False
15. False
16. True
17. Both a and b

Answers to Sample Test Questions from Text Chapter 41 (Marine and Freshwater Environments)

1. Milligrams per liter
2. True
3. Increase
4. Decrease
5. Oxygen
6. Dry weight of filtered biomass
7. False
8. Green sulfur bacteria

9. True
10. Biological
11. Protozoan
12. Enteric bacteria
13. False

Answers to Chapter 40 Sample Test Questions on Microorganisms as Components of Ecosystems

The questions in this test bank have been rated for difficulty. The ratings run from 1 (for the easiest) to 8 (for the most difficult).

#	Answer/Diff		#	Answer/ Diff
1	A / 2		31	C / 3
2	C / 2		32	B / 6
3	B / 3		33	C / 5
4	D / 3		34	C / 4
5	B / 2		35	A / 4
6	C / 3		36	T / 7
7	B / 2		37	F / 6
8	B / 4		38	F / 4
9	C / 5		39	F / 3
10	C / 6		40	F / 3
11	C / 4		41	T / 4
12	D / 5		42	F / 2
13	D / 5		43	F / 2
14	B / 5		44	F / 4
15	A / 4		45	T / 5
16	A / 3		46	F / 3
17	A / 7		47	F / 3
18	D / 6		48	F / 3
19	B / 5		49	F / 4
20	C / 6		50	F / 3
21	B / 5		51	T / 4
22	A / 4		52	F / 5
23	D / 5		53	T / 5
24	A / 4		54	F / 5
25	B / 5		55	T / 3
26	B / 5		56	C / 7
27	B / 6		57	C / 3
28	C / 6		58	T / 7
29	D / 7		59	C / 7
30	C / 8		60	D / 6

Microorganisms in the Environment
Program 9
Microbial Control

*The Lord created medicines out of the Earth, and
the sensible will not despise them.*
—Ecclesiasticus 38:4

One of microbiology's fifty most significant events in the past 125 years.

1935

In the above year, **Gerhard J. Domagk** used a chemically synthesized antimetabolite, Prontosil, to kill streptococci in mice. It was later shown that Prontosil was hydrolyzed in vivo to an active compound, sulfanilamide. One of the first patients to be treated with Prontosil was Domagk's daughter, who had a streptococcal infection that was unresponsive to other treatments. Domagk was awarded the Nobel Prize in Physiology or Medicine in 1939.

Student Assignments

- **Watch**   Video 9: Microbial Control
- **Read**   Chapter 7: Control of Microorganisms by Physical and Chemical Agents,
    Chapter 33: Antimicrobial Chemotherapy, and
    Chapter 43: Microbiology of Food in *Microbiology*, 4th ed., Prescott/Harley/Klein,
    McGraw-Hill.
- **Lab Experiments** from *Laboratory Exercises in Microbiology*, 4th ed., Harley/Prescott,
    McGraw-Hill.
    Exercise 42: Temperature
    Exercise 43: pH
    Exercise 44: Osmotic Pressure
    Exercise 45: Desiccation
    Exercise 46: Radiation (Ultraviolet Light)
    Exercise 47: The Effects of Chemical Agents on Bacteria I: Disinfectants
    Exercise 48: The Effects of Chemical Agents on Bacteria II: Antimicrobial Agents
    Exercise 54: Bacterial Count of a Food Product
    Exercise 55: Coliform Analysis of a Food Product

Lesson Overview

Video 9: *Microbial Control*
In certain situations, microbial control is a necessity. For instance, our food system requires sanitary conditions and hospitals require sterilization techniques. Here we see options available for various levels of microbial control.

Chapter 7: Control of Microorganisms by Physical and Chemical Agents
This chapter focuses on the control and the destruction of microorganisms by physical and chemical agents. This is a topic of great importance, because microorganisms may have deleterious effects, such as food spoilage and disease. It is therefore essential to be able to kill or remove microorganisms from certain environments in order to minimize their harmful effects.

Chapter 33: Antimicrobial Chemotherapy
The control or the destruction of microorganisms that reside within the bodies of humans and other animals is of tremendous importance. This chapter introduces the principles of chemotherapy, and discusses the ideal characteristics for successful chemotherapeutic agents (including the concept of selectively damaging the target microorganism while minimizing damage to the host). The chapter also presents characteristics of some commonly used antibacterial, antifungal, and antiviral drugs.

Chapter 43: Microbiology of Food
This chapter discusses the microorganisms associated with foods. Some of these microorganisms are used in the production of certain foods. Others are disease-causing organisms that are transmitted via foods. Still others are associated with food spoilage. Therefore, the entire sequence of food handling from the producer to the final consumer must be monitored carefully for the presence and activity of microorganisms.

Learning Objectives

After successfully completing this unit, the student will be able to:

- compare and contrast the processes of disinfection, sanitization, antisepsis, and sterilization
- compare the difficulties encountered when trying to kill endospores with those encountered when trying to kill vegetative cells
- discuss the exponential pattern of microbial death
- discuss the influence of environmental factors on the effectiveness of various agents used to control microbial populations
- discuss the uses and limitations of various physical and chemical agents used to control microbial populations
- describe the procedures used to evaluate the effectiveness of various antimicrobial agents
- discuss the various ways in which antimicrobial agents can damage pathogens while causing minimal damage to the host
- discuss the various factors that influence the effectiveness of a chemotherapeutic agent
- discuss the increasingly serious problem of drug-resistant pathogens
- discuss the increasing demand for and availability of antifungal and antiviral agents

- discuss the interaction of intrinsic (food-related) and extrinsic (environmental) factors with the microbial communities associated with foods and the successional changes this interaction produces
- describe the various physical, chemical, and biological processes used to preserve foods
- discuss sterilization processes used in the production of foods and how these processes are monitored
- discuss the various diseases that can be transmitted to humans by foods
- differentiate between food infections and food intoxications
- discuss the detection of disease-causing organisms in foods
- describe the fermentation of dairy products, grains, meats, fruits, and vegetables
- discuss the disease-causing chemicals produced by fungi growing in moist corn and grain products
- discuss wine production by alcoholic fermentation of fruit juices or musts

Course Integrator

Chapter 7
Definition of Frequently Used Terms (p. 136)

Microbes in Motion II	Control/Physical & Chemical/Terminology
Web Link	*Microbiology* Online Learning Center: http://www.mhhe.com/biosci/cellmicro/prescott/
Video *Unseen Life on Earth*	Unit #9 Microbial Control

The Pattern of Microbial Death (p. 137)

Conditions Influencing the Effectiveness of Antimicrobial Agent Activity (p. 138)

The Use of Physical Methods in Control (p. 138)

Visual Resource Library	The Autoclave Figure 7.3 *Microbes in Motion II* Control: Physical & Chemical/Terminology
Microbes in Motion II	Control: Physical & Chemical/Physical Control
Web Link	Sterilization or Disinfection of Medical Devices: General Principles (from CDC) http://www.cdc.gov/ncidod/diseases/hip/sterilgp.htm
Web Link	Sterilization or Disinfection of Patient-Care Equipment: HIV-Related http://www.cdc.gov/ncidod/diseases/hip/hivsteri.htm

The Use of Chemical Agents in Control (p. 143)

Visual Resource Library	Disinfectants and Antiseptics Figure 7.7
Transparency	Disinfectants and Antiseptics Figure 7.7
Microbes in Motion II	Control: Physical & Chemical/Chemical Control
Web Link	CDC Travel Information: Food and Water Precautions http://www.cdc.gov/travel/foodwatr.htm

	/Enterobacteriaceae
	/Spirillaceae
Microbes in Motion II	Gram Positive Bacteria/Gram Positive Bacilli
	/Listeria
	/Bacillus
	/Spore Forming Anaerobes

Microbiology of Fermented Foods (p. 920)

Visual Resource Library	Wine Making Figure 43.16a
Transparency	Wine Making Figure 43.16
Visual Resource Library	Producing Beer Figure 43.17
Transparency	Producing Beer Figure 43.17

Microorganisms as Sources of Food (p. 928)

Microbial terms Used in Video 9 in Order of Appearance with Page Indexing to the Textbook (*Microbiology*, 4th ed., Prescott/Harley/Klein, McGraw-Hill)

disinfectant 147	*Clostridium* 493	droplet spread 729
antibiotics 683	*Salmonella* 477	airborne spread 729
antisepsis 137	*Streptococcus*	colony 107
infection 582	*pyogenes* 489	typhoid 724
sterilization 136	endospores 66	malaria 824
antimicrobial 683	autoclave 139	
	handwashing 700	

Web Links Related to Video 9 (Microbial Control) and Readings in Chapter 7 (Control of Microorganisms by Physical and Chemical Agents), Chapter 33 (Antimicrobial Chemotherapy), and Chapter 43 (Microbiology of Food) (Go to www.mhhe.com/prescott, Student on Line Learning Center, Double Click on Chapter 7 [then 33 and 43], Double Click on Web Sites)

Chapter 7

Sterilization or Disinfection of Medical Devices: General principles. From the Centers for Disease Control
http://www.cdc.gov/ncidod/hip/sterile/sterilgp.htm

Sterilization or Disinfection of Patient-Care Equipment: HIV Related
http://www.cdc.gov/ncidod/hip/sterile/hivsteri.htm

Food and Water Precautions and Traveler's Diarrhea. CDC travel information.
http://www.cdc.gov/travel/foodwatr.htm

Bacteria and Other Microorganisms in Household Water
http://www.cdc.gov/niosh/nasd/docs4/va98020.html

Cleaning and Disinfection of the Child-care Environment
http://www.cdc.gov/ncidod/hip/abc/practic9.htm

Chapter 33

Antiretroviral Drugs Approved by the FDA for HIV
http://www.pedid.uthscsa.edu/014.htm

Antiviral Drugs for Influenza: The Flu-facts, Prevention, and Treatment
http://www.fda.gov/oashi/aids/virals.html

Fungal testing Laboratory Newsletter
http://pharmacology.tqn.com/library/weekly/bl980931.htm

Journal of Antimicrobial Chemotherapy on Line
http://www3.oup.co.uk/jac/contents

Chapter 43

The "Bad Bug Book". This handbook developed by the Food and Drug Administration, provides basic facts regarding foodborne pathogenic microorganisms and natural toxins.
http://vm.cfsan.fda.gov/~mow/intro.html

Food preservation Page
http://www.marketreport.com/preserver.htm

The Latest Word on Mold: When to Toss and When to Trim
http://www.marketreport.com/mold.htm

International Travelers Health Line
http://www.travelhealthline.com/cicc_supp.html

Microbial Terms Used in Chapter 7 (Control of Microorganisms by Physical and Chemical Agents)

algicide 137
antimicrobial agent 138
antisepsis 137
antiseptics 137
autoclave 139
bacteriocide 137
bacteriostatic 137
D value 139
depth filters 141
detergent 147
disinfectant 137
disinfection 137
F value 139
flash pasteurization 141
fungicide 137
fungistatic 137
germicide 137
iodophor 145
ionizing radiation 142
membrane filters 141
pasteurization 141
phenol coefficient test 147
sanitization 137
thermal death time 133
tyndalization 141
ultraviolet radiation 142
use dilution test 148
viricide 137
z value 139

Microbial Terms Used in Chapter 33 (Antimicrobial Chemotherapy)

acyclovir 693
amantidine 693
amphotericin B 692
antibiotic 678
antimetabolites 684
azidothymidine (AZT) 694
broad-spectrum drugs 680
cephalosporins 688
chemotherapeutic agent 678
chloramphenicol 689
cidal 681
dilution susceptibility Tests 681
erythromyocin 689
grisofulvin 692
HIV protease inhibitors 694
interferon 695
Kirby-Bauer method 681
macrolide antibiotic 689
narrow spectrum drugs 680
nystatin 692
parenteral route 684
penicillinase 686
penicillins 686
plasmids 690
quinolones 686
R plasmids 691
selective toxicity 679
side effects 680
static 681
streptomycin 689
structural analogues 684
sulfonamide 684
superinfection 691
tetracycline 689
therapeutic index 680
zidovudine 694

Microbial Terms Used in Chapter 43 (Microbiology of Foods)

aflatoxin 910
bottom yeast 925
canning 914
enology 924
ergotism 910
food-borne infection
  916
food intoxication
917
fumonisins 911
lagered 925

malt 925
mash 925
mashing 924
must 924
pasteurization 914
pitched 925
probe 99
putrefaction 909

racking 924
radapperization 916
silages 928
sour mash 926
starter culture 920
wine vinegar 925
wort 924
zxerophilic
microorganisms
  909

Sample Test Questions from Video 9 (Microbial Control)

1. Which of the following can be used to control microorganisms?
   a. Disinfectants
   b. Antiseptics
   c. Antibiotics
   d. a, b, and c

2. Contamination control is a health issue.
   a. True
   b. False

3. Canning is one of the oldest and most reliable ways humans have controlled microbes.
   a. True
   b. False

4. Which of the following organisms might get into a canned product?
   a. *Salmonella*
   b. *Clostridium*
   c. *Plasmodium*
   d. *Aspergillus*
   e. Both a and b

5. During the American Civil War, more people died from infections than were killed in battle.
   a. True
   b. False

6. In the early 20th century, were there cures for cholera, tuberculosis, and typhoid?
   a. Yes
   b. No

7. Who discovered the first antibiotic, penicillin? ___________

8. Penicillin first became available to the general public in the
   a. 1920s
   b. 1930s
   c. 1940s
   d. 1950s

9. Which of the following diseases are practically extinct today?
   a. Smallpox
   a. Measles
   b. Polio
   c. All of these

10. Viruses are still a major threat to humans since there are very few anti-viral agents.
    a. True
    b. False

Sample Test Questions from Text Chapter 7 (Control of Microorganisms by Physical and Chemical Agents) Web Site

1. Methods for microbial destruction are used for the control of food spoilage and disease.
   O True
   O False

2. Bacterial endospores can be killed by
   O chemical agents
   O some gases and heat
   O all of the above

3. The process by which all living cells are destroyed is called
   O disinfection
   O sterilization
   O sanitization

4. The process by which disease-causing microorganisms are killed is called
   O disinfection
   O sterilization
   O sanitization

5. The process by which a microbial population is reduced to safe levels is called
   - O   disinfection
   - O   sterilization
   - O   sanitization

6. Chemical agents applied to tissue to prevent infection are called
   - O   antiseptics
   - O   disinfectants
   - O   germicides

7. A chemical that inhibits fungal growth is called
   - O   fungicidal
   - O   fungistatic

8. An effective physical means to remove or destroy microorganisms is
   - O   heat
   - O   filtration
   - O   ultraviolet radiation
   - O   ionizing radiation
   - O   all of the above

9. Fungal spores can survive hours of boiling.
   - O   True
   - O   False

10. The decimal reduction time ("D value") refers to the shortest time needed to
    - O   kill all the microorganisms in a sample
    - O   kill half of microorganisms in a sample
    - O   kill 90% of the organisms in a sample

11. Moist heat sterilization must be carried out at temperatures of 100 degrees C for 5 minutes in order to kill bacterial endospores.
    - O   True
    - O   False

12. An autoclave is used for
    - O   moist heating under pressure
    - O   high-temperature, short-term heating
    - O   dry heat sterilization

13. Filtration is an excellent way to reduce microbial populations in solutions. To be effective, membrane filters must have pores about 1mm in diameter.
    - O   True
    - O   False

14. Ultraviolet light of 260 nm is lethal to microorganisms.
   - O   True
   - O   False

15. Pasteurization refers to the process of
   - O   moist heating under pressure
   - O   high-temperature, short-term heating
   - O   dry heat sterilization

16. Ionizing radiation (Gamma radiation) from a cobalt 60 source
   - O   is an excellent sterilizing agent
   - O   penetrates deep into objects
   - O   all of the above

Sample Test Questions from Text Chapter 33 (Antimicrobial Chemotherapy) Web Site

1. Treatment of fungal infections is more difficult than treatment of bacterial infections because of the greater similarity between the fungi and their hosts.
   - O   True
   - O   False

2. A drug that disrupts a microbial function not found in animal cells usually has a higher therapeutic index.
   - O   True
   - O   False

3. A drug that disrupts a microbial function not found in animal cells usually has a higher selective toxicity.
   - O   True
   - O   False

4. The modern era of chemotherapy began with Alexander Fleming's work on drugs against African sleeping sickness and syphilis.
   - O   True
   - O   False

5 Which of the following is a desirable general characteristic of antimicrobial drugs?
   selective toxicity
   - O   broad-spectrum of activity
   - O   must be bactericidal not bacteriostatic
   - O   all of these are desirable general characteristics

6. The lowest concentration of an antibiotic resulting in no growth is the
   - O minimal inhibitory concentration (MIC).
   - O minimal lethal concentration (MLC).
   - O 50% inhibitory dose.

7. In order to determine the minimal lethal concentration (MLC), the culture is first inoculated into medium containing the drug then subcultured into medium containing
   - O the same amount of drug.
   - O no drug.
   - O an increased amount of drug.
   - O a decreased amount of drug (but not zero).

8. In order to be effective, the concentration of a drug in the body must be
   - O higher than the pathogen's minimal inhibitors concentration (MIC).
   - O lower than the pathogen's MIC.

9. Which of the following affects the concentration of a drug in the body?
   - O amount administered
   - O how administered
   - O speed of uptake
   - O rate of elimination from the body
   - O all of these.

10. Which of the following do(es) not inhibit protein synthesis?
    - O tetracycline
    - O streptomycin
    - O erythromycin
    - O penicillin

11. Drugs can be excluded from a cell by
    - O an inability to bind to the cell.
    - O an inability to penetrate the cell.
    - O either an inability to bind to the cell or an inability to penetrate the cell.

12. Genes for drug resistance are located
    - O on the chromosome.
    - O on plasmids.
    - O either on the chromosome or on plasmids.

Sample Test Questions from Text Chapter 43 (Microbiology of Foods) Web Site

1. Foods provide an ideal environment for microbial growth.
   - O True
   - O False

2. Food-borne infections or food intoxications are usually transmitted by the fecal-oral route, with food acting as an intermediary agent.
   O   True
   O   False

3. *E. coli* is a human health concern because of the toxins it produces.
   O   True
   O   False

4. Many spices have antimicrobial properties.
   O   True
   O   False

5. Decreasing water availability by the addition of ________________ controls microbial growth in food.
   O   sugar
   O   salt
   O   both sugar and salt

6. The first person to use heating processes to preserve foods was
   O   Louis Pasteur
   O   Robert Koch
   O   Nicholas Appert

7. A test organism used to measure sterilization efficiency is
   O   *Bacillus* stearothermophilus
   O   *Clostridium botulinum*
   O   *Escherichia coli*

8. Which chemical is used to preserve meat products?
   O   ethylene oxide
   O   propionate
   O   nitrite

9. *Salmonella* causes food-borne infections.
   O   True
   O   False

10. *Campylobacter*, a common bacteria in many red meats and poultry products, can be eliminated by proper
   O   refrigeration
   O   heating

11. Ethylene oxide is often used to control disease-causing organisms in spices.
   O   True
   O   False

12. An organism used to produce vinegar from wine is
    - O Acetobacter
    - O *Lactobacillus*
    - O *Pseudomonas*
    - O *Bacillus*

13. Pasteurization
    - O reduces the total microbial population.
    - O usually eliminates all disease-causing organisms.
    - O both reduces the total microbial population and usually eliminates all disease-causing organisms.

14. Which of the following helps preserve foods by controlling microbial access to water?
    - O drying
    - O addition of salt
    - O addition of sugar
    - O all of these.

15. A limitation of irradiation to sterilize foods is that it will only work on
    - O moist foods
    - O dry foods

Chapter 7 Sample Test Questions on Control of Microorganisms by Physical and Chemical Agents

1. The destruction or removal of all viable organisms is called
    a. disinfection.
    b. antisepsis.
    c. sterilization.
    d. sanitization.

2. The reduction of the microbial population to safe levels is called
    a. disinfection.
    b. antisepsis.
    c. sterilization.
    d. sanitization.

3. The prevention of infection caused by microorganisms is called
    a. disinfection.
    b. antisepsis.
    c. sterilization.
    d. sanitization.

4. The killing of vegetative forms of pathogenic microorganisms (usually on inanimate objects) is called
   a. disinfection.
   b. antisepsis.
   c. sterilization.
   d. sanitization.

5. Microbial populations are reduced by ___________ when exposed to a lethal agent.
   a. a constant number
   b. a constant fraction
   c. a continuously increasing number
   d. a variable number

6. Which of the following influence(s) the efficiency of an antimicrobial agent?
   a. concentration of the agent
   b. duration of exposure
   c. temperature
   d. All of these.

7. Which of the following environmental factors generally do(es) not have an impact on the efficiency of an antimicrobial agent?
   a. pH
   b. viscosity
   c. concentration of organic matter
   d. All of these have an impact on the efficiency of an antimicrobial agent.

8. Which of the following is not a method of heat sterilization?
   a. autoclaving
   b. tyndallization
   c. pasteurization
   d. All of these are methods of heat sterilization.

9. Which of the following is MOST effective against resistant endospores?
   a. autoclaving
   b. tyndallization
   c. pasteurization
   d. All of these are very effective against resistant endospores.

10. Which of the following is not a form of ionizing radiation?
    a. ultraviolet (UV) radiation
    b. X radiation
    c. gamma radiation
    d. All of these are forms of ionizing radiation.

11. Which of the following is (are) a sterilizing gas(es) that can be used on heat-sensitive materials?
    a. ethylene oxide
    b. betapropiolactone
    c. both ethylene oxide and betapropiolactone
    d. neither ethylene oxide nor betapropiolactone

12. A(n) __________ is a chemical that can be used to sterilize materials.
    a. oxidant
    b. disinfectant
    c. antiseptic
    d. sterilant

13. Repeated cycles of exposure to elevated temperatures followed by incubation at 37°C in order to kill heat resistant endospores are called _________.
    a. tyndallization
    b. fractional steam sterilization
    c. Both a and b are correct.
    d. Neither a nor b is correct.

14. The time required to kill 90% of the microorganisms or spores in a sample at a specified temperature is the
    a. thermal death time (TDT).
    b. thermal death point (TDP).
    c. decimal reduction time (D value).
    d. z value.

15. The shortest period of time needed to kill all organisms in a sample at a specified temperature is the
    a. thermal death time (TDT).
    b. thermal death point (TDP).
    c. decimal reduction time (D value).
    d. z value.

16. The increase in temperature needed to reduce the decimal reduction time to 10% of its value is the
    a. thermal death time (TDT).
    b. thermal death point (TDP).
    c. decimal reduction time (D value).
    d. z value.

17. Organic molecules that serve as wetting agents and emulsifiers and can be used as disinfectants if they are cationic are called
    a. sterilants.
    b. oxidants.
    c. detergents.
    d. soaps.

18. Sterilization is accomplished by __________ all viable microorganisms.
    a. killing
    b. removing
    c. either killing or removing
    d. neither killing nor removing

19. Moist heat sterilizes by
    a. degrading nucleic acids.
    b. denaturing proteins.
    c. disrupting membranes.
    d. All of these.

20. Which of the following is not a function of pasteurization?
    a. killing pathogenic microorganisms
    b. reducing the total microbial population
    c. increasing the shelf life of the product
    d. All of these are functions of pasteurization.

21. The disinfectant of choice for municipal water supplies is __________.
    a. fluorine
    b. chlorine
    c. either fluorine or chlorine
    d. neither fluorine nor chlorine

22. Disinfectants are generally regulated by the
    a. Food and Drug Administration.
    b. Environmental Protection Agency.
    c. either Food and Drug Administration or Environmental Protection Agency.
    d. neither Food and Drug Administration nor Environmental Protection Agency.

23. Antiseptics are generally regulated by the
    a. Food and Drug Administration.
    b. Environmental Protection Agency.
    c. either Food and Drug Administration or Environmental Protection Agency.
    d. neither Food and Drug Administration nor Environmental Protection Agency.

24. An agent that specifically kills fungi but not other kinds of microorganisms is best described as a
    a. fungistatic agent.
    b. germistatic agent.
    c. fungicidal agent.
    d. germicidal agent.

25. Which of the following generally do(es) not contribute to resistance to killing?
    a. formation/existence of endospores
    b. increased age of the culture
    c. inherent/genetic differences among organisms
    d. All of these may contribute to resistance to killing.

26. Depth filters sterilize materials passing through them by
    a. exclusion of materials too large to penetrate the filter.
    b. entrapment of material in the channels of the filter.
    c. both exclusion of materials too large to penetrate the filter and entrapment of material in the channels of the filter.
    d. neither exclusion of materials too large to penetrate the filter nor entrapment of material in the channels of the filter.

27. When antiseptics and disinfectants are compared, antiseptics are generally
    a. less toxic.
    b. more toxic.
    c. equally as toxic.
    d. unpredictable in toxicity.

28. Which of the following is true about filtration?
    a. It can be used only to remove organisms from liquids.
    b. It does not truly sterilize because it removes rather than kills microorganisms.
    c. Both of these answers are correct.
    d. Neither of these answers is correct.

29. When comparing dry heat and moist heat sterilization, dry heat is
    a. faster.
    b. slower.
    c. equally as fast.
    d. sometimes faster but sometimes slower.

30. Which will require a longer time to kill?
    a. a larger population of microorganisms
    b. a smaller population of microorganisms
    c. Killing will be equally as rapid in a large or a small microbial population.
    d. There is no way to predict which will require a longer kill time.

31. Although heavy metals are no longer widely used as germicides, __________ is still used to prevent ophthalmic gonorrhea in newborns.
    a. lead acetate
    b. copper sulfate
    c. silver nitrate
    d. mercuric chloride

32. Iodine can be complexed with an organic carrier to form water-soluble, stable complexes called __________, which release iodine slowly and eliminate most problems associated with iodine use.
    a. iodides
    b. tinctures
    c. iodophores
    d. iodochromes

33. The two most important alcohol germicides are __________ and __________.
    a. ethanol; methanol
    b. ethanol; isopropanol
    c. methanol; butanol
    d. methanol; isopropanol

34. Two methods of pasteurization are __________ and __________.
    a. 63°C for 30 minutes; 72°C for 15 minutes
    b. 63°C for 30 minutes; 72°C for 15 seconds
    c. 63°C for 30 seconds; 72°C for 15 seconds
    d. 63°C for 15 minutes; 72°C for 30 seconds

35. The process of heating milk products to 72°C for 15 seconds in order to reduce the microbial population is called __________ pasteurization.
    a. flash
    b. fast
    c. ultra fast
    d. None of these are correct.

36. Which of the following represents the best definition for microbial death?
    a. the organism will not grow on minimal medium
    b. the organism will not grow on a medium that normally supports its growth
    c. the organism no longer retains its original shape and structures
    d. None of these adequately describe microbial death.

37. Which of the following is not a reason for studying methods of destroying microorganisms?
    a. it makes possible microbiological research
    b. preservation of food
    c. prevention of disease
    d. All of these are reasons for studying methods of destroying microorganisms.

38. An agent that kills bacteria is referred to as __________.

39. An agent that prevents the growth of bacteria without causing irreversible damage to the bacteria is referred to as __________.

40. Before being released for general use, antimicrobial agents should be tested for __________ and __________.

41. The __________ is a measure of disinfectant efficiency in which the disinfectant being evaluated is compared to phenol.

42. The type of filter used in a laminar flow biological safety cabinet is called a(n) __________ filter.

43. T F Ultraviolet radiation is an effective means of sterilization but is limited to surfaces.

44.T F Alcohols are widely used as antiseptics and disinfectants because they are effective against endospores as well as vegetative cells.

45.T F Because filtration removes rather than destroys microorganisms, it does not truly sterilize the materials passing through the filter.

46.T F Beta propiolactone is not as useful as ethylene oxide as a sterilizing agent because it does not penetrate materials as readily as ethylene oxide.

47.T F The usefulness of an antiseptic or disinfectant is based solely on its effectiveness against pathogenic organisms.

48.T F The rate of killing by an antimicrobial agent may slow when the microbial population has been greatly reduced because the remaining population may have a high proportion of resistant organisms.

49.T F Ethylene oxide can be used to sterilize materials in sealed plastic wrap because it can penetrate the wrapping material.

50.T F The phenol coefficient is a direct indication of disinfectant potency during normal use.

51.T F Larger populations generally are killed as rapidly as smaller populations.

52.T F Microorganisms show differential sensitivity to antimicrobial agents.

53.T F Dry heat methods usually require lower temperatures and shorter exposure times than moist heat methods to achieve the same degree of killing because of the drying effects of this form of heat.

54.T F Heavy metals are effective antimicrobial agents but are not widely used because of their high toxicity to humans.

55.T F Quaternary ammonium compounds can be used as skin antiseptics because of their low toxicity.

56.T F Although UV radiation is generally only effective against surface organisms, it can penetrate water to sterilize submerged items.

Chapter 33 Sample Test Questions on Antimicrobial Chemotherapy

1.   Who is generally credited with the discovery of sulfanilamide as a chemotherapeutic agent?
      a. Gerhard Domagk
      b. Jacques Trefouel
      c. Therese Trefouel
      d. All of these.

2. Which of the following is not a desirable general characteristic of antimicrobial drugs?
    a. selective toxicity
    b. broad-spectrum of activity
    c. must be bactericidal not bacteriostatic
    d. All of these are desirable general characteristics of antimicrobial drugs.

3. The lowest concentration of an antibiotic resulting in no growth is the
    a. minimal inhibitory concentration (MIC).
    b. minimal lethal concentration (MLC).
    c. 50% inhibitory dose.
    d. All of these result in no growth because they all mean the same thing.

4. In order to determine the minimal lethal concentration (MLC), the culture is first inoculated into medium containing the drug then subcultured into medium containing
    a. the same amount of drug.
    b. no drug.
    c. an increased amount of drug.
    d. a decreased amount of drug (but not zero).

5. In order to be effective, the concentration of a drug in the body must be
    a. higher than the pathogen's minimal inhibitors concentration (MIC).
    b. lower than the pathogen's MIC.
    c. higher than the pathogen's MLC.
    d. lower than the pathogen's MLC.

6. In addition to the amount of the drug administered, which of the following affects the concentration of a drug in the body?
    a. route of administration
    b. speed of uptake from the site of administration
    c. rate of clearance (elimination) from the body
    d. All of these.

7. Which of the following can limit the usefulness of some drugs in patients?
    a. allergies to the drug
    b. superinfection by drug-resistant microflora
    c. deleterious effects on the normal microflora of the body
    d. All of these.

8. Which of the following do(es) not inhibit protein synthesis?
    a. tetracycline
    b. streptomycin
    c. erythromycin
    d. penicillin

9. Drugs can be excluded from a cell by
   a. an inability to bind to the cell.
   b. an inability to penetrate the cell.
   c. either an inability to bind to the cell or an inability to penetrate the cell.
   d. neither an inability to bind to the cell nor an inability to penetrate the cell.

10. Genes for drug resistance are located
    a. on the chromosome.
    b. on plasmids.
    c. either on the chromosome or on plasmids.
    d. neither on the chromosome nor on plasmids.

11. Which of the following is (are) normally treated by topical application of antifungal drugs?
    a. systemic mycoses
    b. superficial mycoses
    c. both systemic and superficial mycoses
    d. neither systemic nor superficial mycoses

12. Which of the following is the major reason why it has been difficult to treat viral infections with chemotherapeutic agents?
    a. Viruses resemble their hosts and therefore offer no selective point of attack.
    b. Viruses use the metabolic machinery of their hosts and therefore offer no selective point of attack.
    c. Viruses have no metabolism and therefore offer no selective point of attack.
    d. Actually, viruses are not difficult to treat with chemotherapeutic agents.

13. The ratio of toxic dose to therapeutic dose is called the
    a. toxicity index.
    b. phenol coefficient.
    c. therapeutic index.
    d. susceptibility quotient.

14. Chemotherapeutic agents that are natural products of microorganisms are most specifically referred to as __________.
    a. antimicrobial agents
    b. synthetic drugs
    c. antibiotics
    d. semisynthetic drugs

15. Chemotherapeutic agents that are artificially produced are most specifically referred to as

    __________.
    a. antimicrobial agents
    b. synthetic drugs
    c. antibiotics
    d. semisynthetic drugs

16. Chemotherapeutic agents that are chemically modified natural products of microorganisms are most specifically referred to as __________.
    a. antimicrobial agents
    b. synthetic drugs
    c. antibiotics
    d. semisynthetic drugs

17. A plasmid bearing one or more antibiotic resistance genes is a(n)
    a. F plasmid.
    b. A plasmid.
    c. R factor.
    d. D factor.

18. The use of arsenic compound Salvarsan as a treatment for syphilis is credited to
    a. Fleming.
    b. Ehrlich.
    c. Waksman.
    d. Chain.

19. Who of the following is (are) credited with the discovery of the   therapeutic value of penicillin?
    a. Howard Florey
    b. Ernst Chain
    c. both Howard Florey and Ernst Chain
    d. neither Howard Florey nor Ernst Chain

20. His discovery of streptomycin stimulated an intense search for other antibiotics.
    a. Waksman
    b. Ehrlich
    c. Fleming
    d. Pasteur

21. Consider an antibacterial drug. In which of the following cases would the action of the drug be considered primarily bacteriostatic?
    a. The minimal lethal concentration (MLC) is equal to the minimal inhibitory concentration (MIC).
    b. The MLC is lower than the MIC.
    c. The MLC is 2-4 times higher than the MIC.
    d. The MLC is 10-20 times higher than the MIC.

22. The most selective antibacterial agents are those that interfere with bacterial cell wall synthesis. This is because
    a. bacterial cell walls have a unique structure not found in eucaryotic cells.
    b. bacterial cell wall synthesis is easily inhibited whereas animal cell wall synthesis is more resistant to the actions of the drugs.
    c. animal cells do not take up the drugs.
    d. animal cells inactivate the drugs before they can do any damage.

23. Combinations of drugs that are more effective because each drug enhances the effects of the others is referred to as
    a. multiplicity.
    b. synergism.
    c. complementation.
    d. additivity.

24. Which of the following is (are) used by microorganisms to become resistant to a particular drug?
    a. enzymatic inactivation of the drug
    b. exclusion of the drug from the cell
    c. an alternate metabolic pathway that bypasses the drug-sensitive step
    d. All of these.

25. Development of drug resistance is encouraged by
    a. use of a high enough concentration of the drug to destroy any spontaneous resistant mutants that might arise.
    b. use of two drugs simultaneously with the hope that each will prevent the emergence of resistance to the other.
    c. avoidance of indiscriminate use of drugs.
    d. All of these.

26. It is difficult to treat central nervous system infections because
    a. the organisms causing such infections are usually resistant to most drugs.
    b. the drugs cannot readily penetrate the blood-brain barrier.
    c. Both of these answers are correct.
    d. Neither of these answers is correct.

27. The fungus Candida Albicans
    a. is normally present in various parts of the body.
    b. usually causes a problem only when bacterial competition is eliminated by antibiotic treatment.
    c. Both of these answers are correct.
    d. Neither of these answers is correct.

28. Isoniazid is a narrow-spectrum antibiotic that
    a. is not useful as a chemotherapeutic agent.
    b. is useful because it is one of the few drugs effective against Mycobacterium Tuberculosis.
    c. may be useful but has never been tested.
    d. None of these.

29. Drugs with highly toxic side effects
    a. are never used.
    b. are sometimes used in life-threatening situations in which suitable alternatives are not available.
    c. are used in combination with other drugs that counteract the side effects.
    d. are used only in research situations.

30. Sulfonamides and other drugs that inhibit folic acid synthesis have a high therapeutic index because
    a. humans do not use folic acid.
    b. humans do not synthesize folic acid but obtain it in their diets.
    c. humans synthesize folic acid by a process that is not sensitive to sulfonamides.
    d. None of these.

31. Antibiotics that are given by injection are said to have a(n) __________ route of administration.
    a. oral
    b. venal
    c. parenteral
    d. topical

32. Sulfonamides inhibit the production of folic acid and, therefore, inhibit   the synthesis of __________, which are needed for nucleic acid synthesis.
    a. purines
    b. pyrimidines
    c. both purines and pyrimidines
    d. neither purines nor pyrimidines

33. Which of the following is likely to have the most toxic side effects to humans?
    a. inhibitors of cell wall synthesis
    b. inhibitors of protein synthesis
    c. disrupters of cell membrane structure
    d. inhibitors of DNA synthesis

34. Which of the following is (are) true about cephalosporins?
    a. They, like penicillin, inhibit bacterial cell wall synthesis.
    b. They have a broader spectrum of activity than penicillin.
    c. They can be given to patients with penicillin allergies.
    d. All of these are true.

35. Penicillin was first discovered by __________ and then the knowledge was lost for 50 years until rediscovered by a librarian.
    a. Fleming
    b. Florey and Chain
    c. Duchesne
    d. Jenner

36. Who is usually credited with the discovery of penicillin even though the discovery was not followed up?
    a. Fleming
    b. Florey and Chain
    c. Duchesne
    d. Jenner

37. Who, working from published observations, demonstrated the effectiveness of penicillin?
    a. Fleming
    b. Florey and Chain
    c. Duchesne
    d. Jenner

38. Which of the following is not a useful mechanism of action for an antimicrobial drug?
    a. inhibition of cell wall synthesis
    b. inhibition of protein synthesis
    c. disruption of cell membranes
    d. All of these are useful mechanisms of action for antimicrobial drugs.

39. Which of the following is not true about drug side effects?
    a. they may result from drug inhibition of the same process in the host as in the target cell
    b. they may result from drug damage to a different process in the host cell than in the target cell
    c. they usually result from damage to the liver of the host when the liver tries to eliminate the drug from the system
    d. All of these are true about drug side effects.

40. Which of the following inhibits protein synthesis?
    a. tetracycline
    b. erythromycin
    c. both tetracycline and erythromycin
    d. neither tetracycline nor erythromycin

41. Which of the following is useful against viral infections?
    a. acyclovir
    b. idoxuridine
    c. both acyclovir and idoxuridine
    d. neither acyclovir nor idoxuridine

42. The most promising development in the area of antiviral treatment is the use of __________, a naturally produced antiviral substance.

43. __________ are activities of a chemotherapeutic agent that damage the   host either by inhibiting the same process in the host as in the target cell or by damaging other processes.

44. When a chemotherapeutic agent attacks many different pathogens, it is said to be a __________ antibiotic.

45. A(n) __________ is a drug that blocks the function of a metabolic pathway.

46. Many antimetabolites block the functioning of metabolic pathways by acting as __________ inhibitors of key enzymes.

47.TF Much of our modern chemotherapy developed from ancient herbal remedies.

48.TF Penicillin is frequently injected but not normally given orally because it is rapidly degraded in the stomach.

49.TF The effects of most drugs are greater if the microorganisms are actively growing.

50.TF Treatment of fungal infections is more difficult than treatment of bacterial infections because the greater similarity between the fungi and their hosts limits the ability of a drug to have a selective toxicity.

51.TF A drug that disrupts a microbial function not found in animal cells usually has a higher therapeutic index.

52.TF Agents that are static and, therefore, do not kill infectious organisms are not useful as chemotherapeutic agents.

53.TF Protein synthesis inhibitors have a low therapeutic index because they usually cannot discriminate between procaryotic and eucaryotic ribosomes.

54.TF Ampicillin, like penicillin, must be administered by injection because it is rapidly degraded in the stomach and cannot, therefore, be given orally.

55.TF Drug allergies and toxic side effects are synonymous terms used to describe the way in which the usefulness of a drug can be limited.

56.TF In a disk diffusion assay for drug susceptibility, a drug producing a larger clear zone always indicates that the organism is more susceptible to that drug than to one producing a smaller clear zone.

57.TF The minimum lethal concentration (MLC) is the lowest concentration of drug from which organisms fail to recover when removed from the drug.

58. The emergence of new strains that are now resistant to antibiotics that they were once sensitive to is becoming a major problem in
a. bacterial diseases.
b. viral diseases.
c. both bacterial and viral diseases.
d. neither bacterial nor viral diseases.

59. Which of the following affects the size of the clear zone in a disk diffusion test of antimicrobial susceptibility?
a. the initial concentration of the drug
b. the solubility of the drug
c. the diffusion rate of the drug
d. All of the above.

60.TF One way in which organisms may exhibit resistance to a drug is the ability to pump the drug out of the cell immediately after it has entered.

61. The drug level required for the clinical treatment of a particular infection is called the
     a. therapeutic dose.
     b. toxic dose.
     c. therapeutic index.
     d. minimal inhibitory concentration.

62.TF One way that the development of drug resistance in viruses has been minimized has been by the use of a cocktail of several drugs at high doses.

Chapter 43 Sample Test Questions on the Microbiology of Food

1. To control microbial growth in a food, one can decrease water availability by the addition of
    a. sugar.
    b. salts.
    c. both sugar and salts.
    d. neither sugar nor salts.

2.TF Recently bioluminescence methods have been used to detect potential pathogens on utensils and other surfaces.

3. Lysozyme is an important antimicrobial substance found at high levels in which food?
    a. fish
    b. eggs
    c. vegetables
    d. fruits

4. The use of plastic film to wrap meat products can allow increased contact with
    a. oxygen.
    b. helium.
    c. carbon dioxide.
    d. water.

5. The first person to use heating processes to preserve foods was
    a. Louis Pasteur.
    b. Jacque Monod.
    c. Robert Koch.
    d. Nicholas Appert.

6. The D value is the time required to have the population of a test organism decreased by
   __________ at a given temperature.
   a. 10%
   b. 1%
   c. 90%
   d. 99%

7. To assure that foods are essentially free of living organisms, __________ D cycle(s) is usually
   used.
   a. 1
   b. 5
   c. 12
   d. 20

8. The Z value is a measure of the change in the die-off rate of a test organism in response to a
   difference in
   a. time.
   b. temperature.
   c. time and temperature.
   d. neither time nor temperature.

9. A major test organism used to measure sterilization efficiency is
   a. *Bacillus Stearothermophilus.*
   b. *Clostridium Botulinum.*
   c. *Escherichia coli.*
   d. *Pseudomonas Aeruginosa.*

10. The term radappertization is used to describe food preservation by
    a. heat.
    b. chemicals.
    c. UV irradiation.
    d. gamma irradiation.

11. An important radiation-resistant bacterial genus is
    a. *Pseudomonas.*
    b. *Escherichia.*
    c. *Bacteroides.*
    d. *Deinococcus.*

12. A major chemical used to preserve meat products is
    a. ethylene oxide.
    b. propionate.
    c. nitrite.
    d. None of these.

13. A major organism considered to cause food-borne infection is
    a. *Salmonella.*
    b. *Staphylococcus.*
    c. *Pseudomonas.*
    d. *Bacillus.*

14. *Campylobacter*, although common in many red meats and poultry products, can easily by eliminated by proper
    a. mixing and grinding.
    b. addition of herbs.
    c. refrigeration.
    d. heating.

15. *Clostridium botulinum* and *Bacillus cereus* are important disease-causing organisms that are food-borne
    a. infections.
    b. intoxicants.
    c. infections and intoxicants.
    d. neither infections nor intoxicants.

16. Ergotism involves microorganism production of
    a. thermostable enzymes.
    b. hallucinogenic alkaloids.
    c. sweeteners.
    d. flavoring compounds.

17. A major organism involved in the production of ergot is
    a. *Pseudomonas fluorescens.*
    b. *Escherichia coli.*
    c. *Claviceps purpura.*
    d. *Aspergillus niger.*

18. Aflatoxins are planar compounds that can cause mutation by reacting with nucleic acids in a process called
    a. substitution.
    b. rearrangement.
    c. deletion.
    d. intercalation.

19. A major fungal genus involved in aflatoxin production is
    a. *Rhizopus.*
    b. *Aspergillus.*
    c. *Claviceps.*
    d. *Penicillium.*

20. Normal milk spoilage, resulting in a clear liquid, involves how many successional steps?
    a. one
    b. two
    c. three
    d. four

21. To control pathogenic and disease-causing organisms in spices __________ is often used.
    a. autoclaving
    b. UV treatment
    c. ethylene oxide
    d. No treatment is necessary.

22. Which of the following microbial genera is not used in the production of fermented dairy products?
    a. *Streptococcus*
    b. *Lactobacillus*
    c. *Leuconostoc*
    d. All are used.

23. The holes in Swiss cheese are produced by the metabolic activities of which bacterial genus?
    a. *Streptococcus*
    b. *Propionibacterium*
    c. *Leuconostoc*
    d. *Lactobacillus*

24. The science of wine production is called
    a. enology.
    b. rheology.
    c. horology.
    d. eurology.

25. The process of racking, used in the production of wines, involves
    a. depressurizing bottles.
    b. addition of sulfites.
    c. addition of carbon dioxide.
    d. removal of sediments.

26. The production of natural carbon dioxide by yeasts after a wine is bottled results in
    a. sweet wine.
    b. dry wine.
    c. champagne.
    d. brandy.

27. A major organism used to produce vinegar from wine is
    a. *Acetobacter.*
    b. *Lactobacillus.*
    c. *Pseudomonas.*
    d. *Leuconostoc.*

28. The process of vinegar formation is carried out under which conditions?
    a. aerobic
    b. anaerobic
    c. both aerobic and anaerobic
    d. neither aerobic nor anaerobic

29. The process of beer lagering involves
    a. carbonation.
    b. storage.
    c. decarbonation.
    d. None of these.

30. Pitching of a batch of wort for beer production involves
    a. sterilization.
    b. acidification.
    c. carbonation.
    d. inoculation.

31. A microorganism that has been used in the production of sour mash whiskey is
    a. *Leuconostoc mesenteroides*
    b. *Lactobacillus delbrueckii*
    c. *Lactococcus lactis*
    d. *Saccharomyces cerevisiae*

32. Bread products can be spoiled by the growth of Bacillus species that produce
    a. gassy doughs.
    b. nonrising breads.
    c. ropiness.
    d. sourdough bread.

33. The production of sauerkraut involves the natural fermentation of chopped cabbage. To have a successful product, salt is added to a concentration of
    a. 1-1.2%.
    b. 2.2-2.8%.
    c. 5-5.3%.
    d. 10-10.5%.

34. A major yeast genus used as a food source or as a source of single-cell proteins (SCP) is
    a. *Rhodotorula.*
    b. *Spirulina.*
    c. *Candida.*
    d. *Lactobacillus.*

35. A major algal food source used in many parts of the world is
    a. *Spirulina.*
    b. Anabaena.
    c. *Rhizobium.*
    d. *Gymnodinium.*

36. Which fungal genus is added in the final step of Roquefort or blue cheese production?
    a. *Aspergillus niger*
    b. *Cladothrix ruber*
    c. *Penicillium notatum*
    d. None of these.

37. Mashing is a process involving
    a. organic acid utilization.
    b. protein putrefaction.
    c. lipid formation.
    d. carbohydrate hydrolysis.

38.TF Growth of yeasts and molds will be favored in more acidic foods.

39. Sequential growth of microorganisms in which the growth of one organism produces environmental conditions suitable for the subsequent growth of another organism is called ____________.

40.TF Once a food product has been processed by heating, it will be less susceptible to contamination than a food that has not been processed.

41.TF Maintaining foods at a temperature of 41 degrees F (5 degrees C) will only retard but not stop most microbial growth.

42.TF Freezing will result in major decreases in microbial populations present in foods.

43.TF A hypertonic solution contains more salt than a hypotonic solution.

44.TF *Staphylococcus*, once it grows in a food, produces toxins that are not usually inactivated by heating.

45.TF Alcoholic beverages can be produced by the direct fermentations of grains.

46.TF Alcoholic beverages can be produced by the direct fermentation of juices of grapes and other fruits

47. Pasteurization
    a. reduces the total microbial population.
    b. usually eliminates all disease-causing organisms.
    c. both reduces the total microbial population and usually eliminates all disease-causing organisms.
    d. neither reduces the total microbial population nor usually eliminates all disease-causing organisms.

48. Which of the following helps preserve foods by controlling microbial access to water?
    a. drying
    b. addition of salt
    c. addition of sugar
    d. All of these.

49. Irradiation has been used to sterilize certain types of foods. There are however, certain limitations on its effectiveness. One of these is that it will only work on
    a. moist foods.
    b. dry foods.
    c. foods that are not in metal cans.
    d. Actually it will work well on all foods.

50. Which of the following determines whether or not a wine is considered to be dry?
    a. the percentage of alcohol after fermentation
    b. the amount of free sugar after fermentation
    c. the amount of carbon dioxide after fermentation
    d. the type of grape used as a starting material

51. Which of the following can function better with lower water availability?
    a. fungi and yeasts
    b. bacteria
    c. viruses
    d. protozoa

52. Preservation by heating is more effective with
    a. acidic foods.
    b. neutral pH foods.
    c. alkaline foods.
    d. Preservation by heating is equally effective at any pH.

53.TF Some fermented dairy products have been suggested to have beneficial effects including a reduction in the incidence of colon cancer and the minimization of lactose intolerance.

54.TF In all cases, it is necessary for disease-causing microorganisms to grow in a food to result in disease transmission to humans.

55. If food-borne disease transmission requires ingestion of the pathogen, followed by growth of the organism, and release of toxins in the intestine it is referred to as a food-borne __________.

56. If food-borne disease transmission does not require growth of the microorganism after ingestion because the toxic substances are already present in the food as a result of previous growth then it is referred to as a food-borne __________.

57. In breadmaking, yeast growth is usually carried out under __________ conditions which results in more carbon dioxide production and less alcohol accumulation.

58. A food spoilage process called ___________ results in the release of foul-smelling amine compounds.

59. Microorganisms that grow in media with low water availability are called ______________.

60. TF Atmosphere is important in food spoilage; high O2 tends to reduce spoilage while high CO2 tends to increase spoilage.

61. Which of the following is true about the use of immunomagnetic reagents in the food industry:
   a. They can be used to remove and concentrate microorganisms for further study.
   d. They can be used to remove microorganisms to reduce spoilage.
   b. Both a and b are true.
   c. Neither a nor b is true.

62. Which of the following is true about molecular methods of microorganism detection in foods?
   a. They can be used to detect the presence of a single specific pathogen.
   b. They can be used to detect viruses that are not easily cultured.
   c. They can be used to detect slow growing pathogens.
   d. All of the above are correct.

63. Which of the following represents ways in which canned foods may undergo spoilage?
   a. spoilage before canning
   b. as a result of underprocessing during canning
   c. leakage of contaminated water through can seams during cooling
   d. All of the above are correct.

64. TF The polymerase chain reaction (PCR) can detect as few as 10 toxin-producing *Escherichia coli* in a population of 100,000 microorganisms isolated from soft cheese samples.

Answers to Sample Test Questions from Video 9

a.	D	f.	No
b.	True	g.	Flemming
c.	True	h.	C
d.	Yes	i.	D
e.	True	10.	True

Answers to Sample Test Questions from Text Chapter 7 (Control of Microorganisms by Physical and Chemical Agents) Web Site

1. True
2. Some gases and heat
3. Sterilization
4. Disinfection
5. Sanitization
6. Antiseptics
7. Fungistatic
8. All of the above
9. False
10. Kill 90% of the organisms in a sample
11. False
12. Moist heat under pressure
13. False
14. True
15. High-temperature short-term heating
16. All of the above

Answers to Sample Test Questions from Text Chapter 33 (Antimicrobial Chemotherapy) Web Site

1. True
2. True
3. True
4. False
5. All of these are desirable characteristics
6. Minimal inhibitory concentration
7. No drug
8. Higher than the pathogen's minimal inhibitory concentration
9. All of these
10. Penicillin
11. Either an inability to bind to the cell or an inability to penetrate the cell
12. Either on the chromosome or on plasmids

Answers to Sample Test Questions from Text Chapter 43 (Microbiology of Food) Web Site

1. True
2. True
3. True
4. True
5. Both sugar and saalt
6. Nicholas Appert
7. *Bacillus stearothermophilus*
8. Nitrite
9. True
10. Heating
11. True
12. *Acetobacter*
13. Both reduces the total microbial population and usually eliminates all disease-causing organisms
14. All of these
15. Moist foods

Answers to Chapter 7 Sample Test Questions on Control of Microorganisms by Physical and Chemical Agents

The question in this test bank have been rated for difficulty. The ratings run from 1 (for the easiest) to 8 (for the most difficult).

#	Answer / Diff.		#	Answer/Diff
1	C / 3			
2	D / 3		30	A / 4
3	B / 3		31	C / 6
4	A / 3		32	C / 7
5	B / 6		33	B / 5
6	D / 5		34	B / 8
7	D / 6		35	A / 5
8	C / 8		36	B / 5
9	B / 6		37	D / 4
10	A / 7		38	bactericidal / 2
11	C / 6		39	bacteriostatic / 2
12	D / 6		40	effectiveness; safety / 6
13	B / 5		41	phenol coefficient / 5
14	C / 5		42	HEPA (high efficiency particulate air) / 7
15	A / 5		43	T / 5
16	D / 5		44	F / 5
17	C / 6		45	F / 5
18	C / 4		46	T / 7
19	D / 3		47	F / 5
20	D / 5		48	T / 6
21	C / 7		49	T / 8
22	B / 6		50	F / 6
23	A / 6		51	F / 4
24	C / 2		52	T / 5
25	D / 4		53	F / 6
26	C / 6		54	T / 5
27	A / 6		55	T / 6
28	D / 7		56	F / 7
29	B / 6			

Answers to Chapter 33 Sample Test Questions on Antimicrobial Chemotherapy

The question in this test bank have been rated for difficulty. The ratings run from 1 (for the easiest) to 8 (for the most difficult).

#	Answer/Diff	#	Answer/Diff
1	D / 5		
2	C / 8		
3	A / 5	32	C / 4
4	B / 5	33	C / 8
5	A / 5	34	D / 5
6	D / 6	35	C / 6
7	D / 7	36	A / 5
8	D / 3	37	B / 6
9	C / 5	38	D / 4
10	C / 5	39	C / 7
11	B / 4	40	C / 4
12	B / 7	41	C / 6
13	C / 5	42	interferon / 4
14	C / 5	43	Side effects / 8
15	B / 5	44	broad-spectrum / 6
16	D / 5	45	antimetabolite / 8
17	C / 4	46	competitive / 5
18	B / 5	47	T / 2
19	C / 4	48	T / 4
20	A / 5	49	T / 5
21	D / 8	50	T / 6
22	A / 6	51	T / 4
23	B / 7	52	F / 5
24	D / 4	53	F / 7
25	D / 6	54	F / 6
26	B / 6	55	F / 4
27	C / 7	56	F / 7
28	B / 5	57	T / 5
29	B / 6	58	A / 6
30	B / 8	59	D / 7
31	C / 6	60	T / 6
		61	A / 4
		62	T / 5

Answers to Chapter 43 Sample Test Questions on Microbiology of Foods

The questions in this test bank have been ranked for difficulty. The ratings run from 1 (for the easiest) to 8 (for the most difficult).

#	Answer/Diff		#	Answer/Diff
			32.	C/7
			33.	B/5
1.	c/4		34.	C/5
2.	T/6		35.	A/4
3.	B/6		36.	D/7
4.	A/6		37.	D/5
5.	D/5		38.	T/5
6.	C/7		39.	Succession/4
7.	C/5		40.	F/4
8.	B/7		41.	T/4
9.	A/5		42.	F/3
10.	D/7		43.	T/5
11.	D/5		44.	T/5
12.	C/4		45.	F/5
13.	A/5		46.	T/4
14.	D/8		47.	C/5
15.	B/8		48.	D/5
16.	B/6		49.	A/5
17.	C/6		50.	B/6
18.	D/6		51.	A/6
19.	B/4		52.	A/6
20.	D/4		53.	T/6
21.	C/4		54.	F/4
22.	D/5		55.	Infection/4
23.	B/4		56.	Intoxication/4
24.	A/3		57.	Aerobic/4
25.	D/5		58.	Putrefaction/6
26.	C/5		59.	Xerophilic/6
27.	A/5		60.	F/4
28.	A/4		61.	A/5
29.	B/4		62.	D/5
30.	D/4		63.	D/4
31.	B/7			

Microorganisms and Humans
Program 10
Microbial Interactions

> Yeasts, molds, mushroom, mildews, and other fungi
> pervade our world. They work great good and terrible evil.
> Upon them, indeed, hangs the balance of life; for without
> their presence in the cycle of decay and regeneration,
> neither man nor any other living thing could survive.
>
> –Lucy Kavaler

One of microbiology's fifty most significant events in the past 125 years

1941

In the above year, **George Beadle** and **Edward Tatum** jointly publish a paper on their experiments using the fungus *Neurospora crassa* to establish that particular genes are expressed through the action of specific enzymes. The first gene to be identified controlled the synthesis of an enzyme in a series that led to the generation of niacin. This report is the genesis of the "one gene-one enzyme" concept. With Lederberg, Beadle and Tatum were awarded the Nobel prize in Physiology or Medicine in 1958.

Student Assignments

- **Watch**   Video 10: Microbial Interactions
- **Read**   Chapter 25: The Fungi (Eumycota), Slime Molds, and Water Molds and
  Chapter 28: Symbiotic Associations: Commensalism, Mutualism, and the Normal
  Microbiota of the Human Body in *Microbiology*, 4th ed.,
  Prescott/Harley/Klein, McGraw-Hill.
- **Lab Experiments** from *Laboratory Exercises in Microbiology*, 4th ed., Harley/Prescott,
  McGraw-Hill.
  Exercise 73: Fungi I: Yeasts
  Exercise 74: Fungi II: Phycomycetes, Ascomycetes, and Basidiomycetes

Lesson Overview

Video 10: *Microbial Interactions*
There are many symbiotic relationships among microbes and between microbes and higher organisms. Students will examine fundamental examples of these relationships.

Chapter 25: The Fungi (Eumycota), Slime Molds, and Water Molds
This chapter discusses the characteristics of the members of the kingdom *Fungi*. The diversity of these organisms is described, and their ecological and economic impact is discussed. In addition, certain protists the slime molds and water molds, which resemble fungi are also presented in this chapter.

Chapter 28: Symbiotic Associations: Commensalism, Mutualism, and the Normal Microbiota of the Human Body
This chapter discusses the relationships between populations of microorganisms living in close association with each other. The discussion focuses on commensalism and mutualism. Examples of each type of relationship are presented. Additionally, a monitored or gnotobiotic condition in which the identities of all microorganisms present are known is discussed. Finally, the chapter presents an overview of the normal microorganisms associated with the human body.

Learning Objectives

After successfully completing this unit, the student will be able to:

- discuss the distribution of fungi and their roles in the environment
- discuss the morphological characteristics of the fungi
- describe the external digestion of organic matter by fungi
- explain the formation of both asexual and sexual spores for reproduction
- discuss the five major types of organisms in this kingdom (zygomycetes, ascomycetes, basidiomycetes, deuteromycetes, and chytrids) and the basis upon which fungi are assigned to these categories
- discuss the slime molds and water molds, and their resemblance to fungi, even though they are phylogenetically distinct
- discuss symbiosis, commensalism, and mutualism
- give some examples of endosymbiosis and ectosymbiosis
- discuss the establishment and maintenance of microbially monitored (gnotobiotic) animal colonies
- discuss the type and distribution of the microbiota (normal microflora) of the human body

Course Integrator

Chapter 25
Distribution (p. 524)

Visual Resource Library	Phylogeny of the Fungi & Molds Diagram 25
Transparency	Phylogeny of the Fungi & Molds Diagram 25
Web Link	*Microbiology* Online Learning Center: (http://www.mhhe.com/biosci/cellmicro/prescott/)
Web Link	Mycological Resources on the Internet: (http://www.keil.ukans.edu/~fungi/)

Web Link	Mycological Society of American Home page: (http://www.erin.utoronto.ca/~w3msa/)

Importance (p. 524)

Structure (p. 524)

Visual Resource Library	Yeast Structure Figure 25.3
Transparency	Yeast Structure Figure 25.3
Visual Resource Library	Hyphal Morphology Figure 25.6
Microbes in Motion II	Mycology/General Eucaryotic Structures /Specific Fungal Features
Web Link	Medical Mycology Research Center - University of Texas Medical Branch (http://fungus.utmb.edu/myco.htm)
Web Link	Mycology Image Library and Diagnostic Key - University of Wisconsin-Madison (http://www.medsch.wisc.edu/medmicro/myco/mycology. html)

Nutrition and Metabolism (p. 527)

Reproduction (p 528)

Visual Resource Library	Rhizopus Life Cycle Figure 25.9
Microbes in Motion II	Mycology/Metabolism & Growth Chapter /Fungal Reproduction

Characteristics of the Fungal Divisions (p. 530)

Visual Resource Library	Ascomycete Life Cycle Figure 25.12
Visual Resource Library	Basidiomycete Life Cycle Figure 25.14
Microbes in Motion II	Mycology/Taxonomy & Nomenclature /Taxonomy-Morphology

Slime Molds and Water Molds (p. 535)

Chapter 28
Types of Symbiosis, Functions and Examples (p. 566)

Visual Resource Library	Tube Worm-Bacteria Symbiosis Figure 28.4 b-d
Transparency	Tube Worm-Bacteria Symbiosis Figure 28.4 b, d
Visual Resource Library	Rumen Biochemistry Figure 28.6a
Microbes in Motion II	Environmental Microbiology/Microbial Ecology/Symbiosis
Microbes in Motion II	Environmental Microbiology/Aquatic Habitats/Vents
Web Link	*Microbiology* Online Learning Center http://www.mhhe.com/biosci/cellmicro/prescott/
Video *Unseen Life on Earth*	Unit #10 Host-Parasite Relations

Gnotobiotic Animals (p. 571)

Distribution of the Normal Microbiotia of the Human Body (p. 573)

Visual Resource Library — Normal Human Microbiota Figure 28.8

Microbes in Motion II — Microbial Pathogenesis/Nonspecific Host Defense /Normal Flora

Microbial Terms Used in Video 10 in Order of Appearance with Page Indexing to the Textbook (*Microbiology*, 4th ed., Prescott/Harley/Klein, McGraw-Hill)

cellulose 570	fermentation 174	termites 567
organic acids 942	methane producers	mutualism 566
rumen 570	571	slime mold 538
cellulases 570	symbiont 566	fungi 530
	Entodinium 562	mycology 523
		methanotrophs 903

Web Links Related to Video 10 (Microbial Interactions) and Readings in Chapter 25 (The Fungi [Eumycota], Slime Molds, and Water Molds) and Chapter 28 (Symbiotic Associations: Commensalism, Mutualism, and the Normal Microbiota of the Human Body) (Go to www.mhhe.com/prescott, Student on Line Learning Center, Double Click on Chapter 25 [and then 28], Double Click on Web Sites

Chapter 25

Fungi Online
http://www.people.cornell.edu/pages/pl18/links.html

The World-wide Web Virtual Library
http://kaw.keil.ukans.edu/~fungi/

The Mycological Society of America
http://www.erin.utoronto.ca/~w3msa/

Chapter 28

Normal Microbial Flora
http://www.pedid.uthscsa.edu/002.htm

International Culture Collection of Arbuscular and Vesicular-Arbuscular Mycorrhizal Fungi
http://invam.caf.wvu.edu

A Look at Mycorrhizal Symbiosis
http://webtutor.tamu.edu/Students/Carpio/project/alook.html

Rhizobium Legume Symbiosis Links and References
http://web.bham.ac.uk/wjw684/dog/links.html

Microbial terms Used in Chapter 25 (The Fungi [Eumycota], Slime Molds, and Water Molds)

antheridium 531
arthroconidia 528
arthrospore 528
ascocarp 531
ascogenous hypha
 531
ascogonium 531
ascomycetes 530
ascospore 529
ascus 531
basidiocarp 534
basidiomycetes 534
basidiospore 529
basidium 534
basidiospore 529
basidium 535
blastospore 529
cellular slime molds
 535
chitin 525
chlamydodospore 529
chytrids 534
coenocytic 526

conidiospore 529
crytococcus 534
dikaryotic stage 529
deuteromycetes 534
ergot 530
ergotism 530
eumycota 523
fungus 523
gametangium 529
hypha 526
mold 526
mycelium 526
mycologists 523
mycology 523
mycosis 523
mycotoxicology 523
myxamoeba 535
oomycetes 537
plasmodial slime
 molds 535

plasmodium 535
progametangium
530
pseudoplasmodium
 535
saprophyte 527
septa 526
septate 526
slime mold 535
sorocarp 535
sorus 535
sporangiospore 529
sporangium 529
swarm cell 535
thallus 524
water mold 535
yeast 526
YM shift 526
zygomycetes 530
zygospore 529

Microbial Terms Used in Chapter 28 (Symbiotic Associations: Commensalism, Mutualism, and the Normal Microbiota of the Human Body)

acne vulgaris 575
comedo 575
commensalism 566
commensal 566

cross-feeding 566
crustose 567
Doderlein's bacillus
 577

ectosymbiosis 566
endosymbiont 569
endosymbiosis 566

Sample Test Questions from Video 10 (Microbial Interactions)

1. Cows are able to digest cellulose because their stomach's contain microbes that produce the enzyme __________ .

2. The relationship between the microbes in the cow's rumen and the cow is termed __________ .

3. The relationship in which two organisms benefit by living together is called __________ .

4. Most plant material in the environment is broken down into carbon compounds by __________ .

5. The science that studies fungi is called __________ .

6. The bacteria that break down methane are called __________ .

7. The specific biochemical process that occurs in the rumen of a cow is called __________ .

8. Chlorinated solvents in the soil can be broken down by __________ .

9. The relationship between the microbes in a termite's gut and the well being of the termite is called __________ .

10. __________ is a general term referring to the microbially mediated decomposition of paper, paint, textiles, hydrocarbons, and other materials.

Sample Test Questions from Text Chapter 25 (The Fungi [Eumycota], Slime Molds, and Water Molds) Web Site

1. Fungi exist mainly as single cells.
   - O   True
   - O   False

2. A mold consists of long, branched threadlike filaments of cells, the hyphae, that form a tangled mass called a
   - O   filament
   - O   mycelium
   - O   sorus

3. Fungi digest insoluble organic matter by secreting exoenzymes, then absorbing the solubilized nutrients.
   - O   True
   - O   False

4. The most important ecological role of fungi is
   - O   producer
   - O   decomposer
   - O   pathogen

5. Fungi digest macromolecules by
   - O   phagocytosis.
   - O   intracellular digestion.
   - O   extracellular enzyme activity.
   - O   active transport.

6. Phytophthora infestans is the cause of a serious plant disease called
   - O   early blight of tomato
   - O   late blight of potato
   - O   apple scab

7. Diseases in animals caused by fungi are called
   - O   fungoses
   - O   mycoses
   - O   cystoses

8. A mycotoxin is a toxic substance produced by a fungus.
   - O   True
   - O   False

276

9. Fungi form associations with which of the following?
   - O plant roots
   - O cyanobacteria
   - O both plant roots and cyanobacteria

10. Fungi are found primarily in which of the following environments?
    - O marine
    - O freshwater
    - O terrestrial

11. The fungi are now separated from plants based on the chemistry of their cell walls.
    - O True
    - O False

12. Fungal spores are usually responsible for the bright colors and fluffy textures of the fungi that produce them.
    - O True
    - O False

13. The mushroom is really the fruiting body of a fungus whose vegetative mycelium inhabits the soil.
    - O True
    - O False

Sample Test Questions from text Chapter 28 (Symbiotic Associations: Commensalism, Mutualism, and the Normal Microbiota of the Human Body) Web Site

1. What is the term for the relationship when two or more different organisms live together in close association?
   - O symbiotic
   - O parasitic
   - O mututalistic

2. In which kind of relationship does one organism benefit while the other is neither harmed or helped?
   - O commensalism
   - O symbiosis
   - O parasitism
   - O mutualism

3. In a mutualistic relationship, both members benefit.
   - O   True
   - O   False

4. Which of the following is an example of commensalism?
   - O   lichens
   - O   microflora on human skin
   - O   nitrogen-fixing Rhizobium on plant roots

5. Most organisms that colonize the human body are
   - O   fungi
   - O   viruses
   - O   bacteria

6. Ectosymbiosis is a relationship in which the two symbiotes remain outside of one another.
   - O   True
   - O   False

7. Which of the following is NOT part of a lichen association?
   - O   Ascomycetes (fungi)
   - O   green algae
   - O   Brown algae
   - O   cyanobacteria

8. In a lichen, the fungal partner
   - O   protects the alga from excess light intensities.
   - O   provides water and minerals to the alga.
   - O   provides a substratum within which the alga grows.
   - O   all of these.

9. In the rumen, food is quickly attacked by the cellulolytic
   - O   anaerobic bacteria
   - O   fungi
   - O   protozoa
   - O   all of these

10. The microorganisms normally associated with a particular tissue are called
    indigenous microbial population.
    - O   microbiota
    - O   microflora
    - O   any of these

11. Microbiota of the skin are most likely to be found
    - O   where there is little moisture
    - O   where the pH of the skin secretions is below 4
    - O   in association with sebaceous glands
    - O   as transient microbiota

12. Under normal circumstances, the microbiota of the skin is kept in check by
    O   a high pH (above 8)
    O   a pH between 4-6
    O   phagocytic activity
    O   a high loss rate

13. In lichens, the fungi send projections of their hyphae across the algal cell wall in order to obtain nutrients from the photosynthetic partner. These projections are called
    O   mycelia
    O   hyphal extensions
    O   haustoria
    O   lichenthropes

14. Which of the following areas of the human body are normally free of microorganisms?
    O   respiratory tract
    O   kidneys
    O   eyes
    O   all of these
    O   just a and b

15. The lungs are protected from microorganisms by
    O   the mucociliary blanket
    O   lysozyme in mucus
    O   phagocytic action of alveolar macrophages
    O   all of these

16. Enterobacteriaceae are found normally in the stomach.
    O   True
    O   False

17. The colon has the largest microbial population in the body.
    O   True
    O   False
    O

18. *Escherichia coli* is the most prevalent bacterium found in the colon.
    O   True
    O   False

19. The oral cavity becomes colonized with microorganisms within hours after birth.
    O   True
    O   False

Chapter 25 Sample Test Questions on The Fungi (Eumycota), Slime Molds, and Water Molds

1.   Fungi
   a.   are decomposers.
   b.   are parasites.
   c.   are useful food sources.
   d.   All of these.

2.   Fungi
   a.   are eucaryotic organisms.
   b.   produce spores.
   c   lack chlorophyll.
   d.   All of these.

3.   The body or vegetative structure of a fungus is called a
   a.   mold.
   b.   thallus.
   c.   hypha.
   d.   yeast.

4.   Which of the following is not true with respect to a yeast?
   a.   contains two or more nuclei
   b.   reproduces either asexually by budding or sexually through spore formation
   c.   is larger than a bacterium
   d.   possesses eucaryotic organelles

5.   The tangled mass or tissue like aggregation in a mold is called a
   a.   hyphae.
   b.   hypha.
   c.   mycelium.
   d.   thallus.

6.   Dimorphism can be advantageous to fungi in that
   a.   it allows them to reproduce more rapidly during adverse environmental conditions.
   b.   it allows them to fend off certain defensive responses.
   c.   it gives them a competitive advantage in obtaining nutrients from the environment over other organisms.
   d.   All of these.

7.   Which of the following is not a medically important dimorphic fungus?
   a.   *Candida albicans*
   b.   *Sporothrix schenckii*
   c.   *Coccidioides capsulatum*
   d.   *Neurospora crassa*

8. Which of the following is true with respect to fungi?
   a. Some are aerobic.
   b. Some are facultatively anaerobic.
   c. Some are obligately anaerobic.
   d. All of the above are true.

9. Asexual reproduction in fungi can be accomplished by
   a. transverse fission of a parent cell into two daughter cells.
   b. hyphal fragmentation.
   c. budding of somatic vegetative cells.
   d. All of these.

10. In fungi, spores produced by budding are called
    a. blastospores.
    b. conidiospores.
    c. sporangiospores.
    d. arthrospores.

11. The common bread mold *Rhizopus solonifer* belongs to the division
    a. *Zygomycota.*
    b. *Ascomycota.*
    c. *Basidiomycota.*
    d. *Deuteromycota.*

12. The ascomycetes are named for their characteristic reproductive structure called a(n)
    a. hypha.
    b. ascus.
    c. thallus.
    d. ascocarp.

13. A basidium
    a. is a fruiting body.
    b. is produced at the tip of hyphae.
    c. is club-shaped, like an ascus.
    d. All of these.

14. In the classical scheme of fungal classification, if a fungus lacks a sexual reproduction phase, or if this phase has not been observed, it is placed within the
    a. *Ascomycota*
    b. *Basidiomycota*
    c. *Deuteromycota*
    d. *Zygomycota*

15. The water molds resemble fungi in
    a. appearance
    b. lifestyle
    c. appearance and lifestyle
    d. neither appearance nor lifestyle

16. The oomycetes
    a. are known as water molds.
    b. do not resemble fungi in appearance.
    c. have cell walls of chitin.
    d. do not contain hyphae.

17. Zygomycetes live on
    a. bread.
    b. organic matter in the soil.
    c. plants.
    d. All of these.

18. Which of the following is not found among the zygomycetes, ascomycetes, and basidiomycetes?
    a. saprophytic nutrition
    b. septate hyphae
    c. flagellated cells
    d. fusion of gametes

19. Which of the following yeasts is not pathogenic to humans?
    a. *Saccharomyces*
    b. *Candida*
    c. *Cryptococcus*
    d. both *Saccharomyces* and *Candida*

20. The hallucinogen LSD is a product of
    a. yeasts.
    b. Fungi Imperfecti.
    c. ascomycetes.
    d. basidiomycetes.

21. Toadstools are good examples of
    a. oomycetes.
    b. ascomycetes.
    c. basidiomycetes.
    d. yeasts.

22. In mushrooms, the sexual spores are produced in a(n)
    a. basidium.
    b. ascus.
    c. sporophyte.
    d. sporangiophore.

23. The smuts and rusts of grain crops are good examples of
    a. deuteromycetes.
    b. basidiomycetes.
    c. oomycetes.
    d. Fungi Imperfecti.

24. Fungi digest macromolecules by
    a. phagocytosis.
    b. intracellular digestion.
    c. extracellular enzyme activity.
    d. active transport.

25. Sexual spores that are flagellated and produced by the chytrids are called ___________.

26. Chytrids do not have true mycelia. Instead they have a multinucleated mass that resembles mycelia.

27. The fruiting body of a fungus produces
    a. gametes.
    b. spores.
    c. rhizoids.
    d. zygotes.

28. In dikaryotic mycelia, there is no possibility of
    a. mitosis.
    b. meiosis.
    c. gene expression.
    d. protein synthesis.

29. When the fungi that never undergo sexual reproduction get together, their watch word could be
    a. "This bud's for you."
    b. "Nobody's perfect."
    c. "Reach out and touch someone."
    d. "Mycelium or your celium."

30. Asexual spores such as conidia develop by the process of
    a. meiosis.
    b. mitosis.
    c. parthenogenesis.
    d. fission.

31. The primary ecological function of fungi is that of
    a. producer.
    b. consumer.
    c. pathogen.
    d. decomposer.

32. *Phytophthora infestans*, the cause of late blight of potato, is a(n)
    a. oomycete.
    b. zygomycete.
    c. basidiomycete.
    d. ascomycete.

33. The myxomycetes are characterized by
    a. the formation of conidia.
    b. the formation of myxoconidia.
    c. the formation of a plasmodium.
    d. All of these.

34. The pseudoplasmodium of the acrasiomycetes differentiates into a
    a. sorocarp.
    b. zygospore.
    c. sporangiophore.
    d. sorocarp and zygospore.

35. The sac fungi are distinguished from all other groups of fungi because their sexual spores are
    a. inside an ascus.
    b. borne on basidia.
    c. zygospores.
    d. motile.

36. Diseases in animals caused by fungi are called
    a. fungoses
    b. mycoses
    c. dictyoses
    d. cystoses

37. A __________ is a toxic substance produced by a fungus.

38. Fungi form associations with which of the following?
    a. plant roots
    b. cyanobacteria
    c. both plant roots and cyanobacteria
    d. neither plant roots nor cyanobacteria

39. Fungi are found primarily in which of the following environments?
    a. marine
    b. freshwater
    c. terrestrial
    d. arboreal

40. In which of the following are the sexual spores haploid?
    a. Zygomycetes
    b. Ascomycetes
    c. Basidiomycetes
    d. Deuteromycetes

41.TF The fungi are now separated from plants based on the chemistry of their cell walls.

42.TF Although mating in yeast strains leads to sexual reproduction, asexual reproduction by budding predominates.

43.TF Like some bacteria, fungal cells secrete enzymes that promote exodigestion after which the released nutrients are transported into the cell across the plasma membrane.

44.TF Fungal spores are usually responsible for the bright colors and fluffy textures of the molds that produce them.

45.TF In the ceonocytes, protoplasmic streaming through hyphae is interrupted by cross walls.

46.TF The germination of myxomycetes produces nonflagellated cells called myxamoebae.

47. Edible truffles are in the division __________, while edible mushrooms are in the division __________.

48. The body of a fungus consists of a __________.

49. Each hypha is bounded by a cell wall of __________.

50. The YM shift occurs when dimorphic fungi switch from the single-celled __________ form to the hyphal __________ form.

51. The heterotropic and absorptive nutrition of the fungi places them in the role of __________.

52. In all fungi, gametes are produced within a single cell, the __________.

53. The __________ is really the fruiting body of a fungus whose vegetative mycelium inhabits the soil.

54. The plasmodial slime molds are called __________.

55. Matching
    1. conidia   a. Found at the tips of special hyphae
    2. zoospores   b. Reproduction in yeast cells
    3. budding   c. Give rise to aquatic fungi

56.   Matching
       1. coenocytes   a. Sac fungi
       2. higher fungi   b. Bread mold
       3. chytridiomycetes   c. Water molds

57.   Matching
       1. ascospores   a. Penetrate bread
       2. rhizoids   b. Basidiomycetes
       3. basidium   c. Ascus

58.   Matching
       1. Irish potato blight   a *Ascomycetes*
       2. black bread mold   b. *Oomycetes*
       3. brewer's yeast   c. *Zygomycetes*

59.   When the cells of cellular slime molds aggregate, they secrete a sheath around the entire mass of cells. This mass, called a __________, then moves, leaving a slime trail behind it.

60.TF In all fungi that have been observed to have a sexual reproduction cycle, male and female gametes are produced on separate hyphae and then must find each other for fertilization to occur.

61.   Self-fertilizing fungi in which male and female gametes are produced on the same mycelia are said to be __________.

62.   Fungi that require outcrossing between different but sexually compatible mycelia are said to be __________.

Chapter 28 Sample Test Questions on Symbiotic Associations: Commensalism, Mutualism, and the Normal Microbiota of the Human Body

1.   Any microorganism that spends a portion of its life associated with another organism of a different species is called a
      a. symbiont.
      b. symbiote.
      c. parasite.
      d. symbiont and symbiote.

2.   Which of the following is not a category of symbiosis?
      a. mutualism
      b. phoresis
      c. commensalism
      d. parasitism

3. In commensalism
    a. both the host and commensal "eat at the same table."
    b. the commensal is metabolically dependent on the host.
    c. the host provides some factor that the commensal needs.
    d. both the host and commensal "eat at the same table" and the commensal is metabolically dependent on the host.
    e. All of these.

4. In mutualism
    a. the commensal is metabolically dependent on the host.
    b. a reciprocal benefit accrues to both partners.
    c. the host is metabolically dependent on the commensal.
    d. All of these.

5. A relationship in which the two symbiotes remain outside of one another is referred to as a(n) ____________.

6. Which of the following is not part of the lichen association?
    a. ascomycetes
    b. green algae
    c. brown algae
    d. cyanobacteria

7. Which type of lichen is compact and appressed to a substratum?
    a. crustose
    b. foliose
    c. fruticose
    d. mycose

8. In a lichen, the fungal partner
    a. protects the alga from excess light intensities.
    b. provides water and minerals to the alga.
    c. provides a substratum within which the alga grows.
    d. All of these.

9. Which of the following may contain zooxanthellae?
    a. jellyfish
    b. sea anemones
    c. corals
    d. All of these.

10. The reason coral reefs are among the most productive and successful ecosystems is due to
    a. the coral-algal mutualistic relationship.
    b. the lichen mutualistic relationship.
    c. commensalism.
    d. All of these.

11. Which of the following is (are) bacteria?
    a. *Photoblepharon*
    b. *Photobacterium*
    c. *Beneckea*
    d. both *Photobacterium* and *Beneckea*

12. The specialized structure in a tube worm in which endosymbiotic, chemolithotrophic bacteria live is called a(n) ___________.

13. In the rumen, food is quickly attacked by the cellulolytic
    a. anaerobic bacteria.
    b. fungi.
    c. protozoa.
    d. All of these.

14. In the rumen, the carbohydrate degraded is fermented to
    a. fatty acids.
    b. carbon dioxide.
    c. methane.
    d. All of these.

15. The microorganisms normally associated with a particular tissue can be referred to as the
    a. indigenous microbial population.
    b. microbiota.
    c. microflora.
    d. All of these.

16. The most numerous bacterium associated with the skin glands is
    a. *Staphylococcus epidermidis.*
    b. *Pityrosporum ovale.*
    c. *Propionibacterium acnes.*
    d. *Pityrosporum orbiculare.*

17. If a comedo becomes plugged, inflammatory lesions called ___________ may result.
    a. papules
    b. pustules
    c. nodules
    d. All of these.

18. The most predominant bacteria found in the nares is (are)
    a. *Staphylococcus aureus.*
    b. *Staphylococcus epidermidis.*
    c. *Propionibacterium acnes.*
    d. *Staphylococcus aureus* and *Staphylococcus epidermidis.*

19. Which of the following would you not find in the nasopharynx?
    a. *Streptococcus peumoniae*
    b. *Neisseria meningitidis*
    c. *Haemophilus influenzae*
    d. *Branhamella catarrhalis*

20. Which of the following would you expect to find in the tonsilar crypts?
    a. *Bacteroides*
    b. *Staphylococcus*
    c. *Streptococcus*
    d. *Neisseria*

21. Which of the following is not a harmless commensal found in the colon?
    a. *Trichomonas hominis*
    b. *Candida albicans*
    c. *Entamoeba hartmanni*
    d. *Endolimax nana*

22. Microbiota of the skin are most likely to be found
    a. where there is little moisture.
    b. where the pH of the skin secretions is below 4.
    c. in association with sebaceous glands.
    d. as transient microbiota.

23. Under normal circumstances, the microbiota of the skin is kept in check by
    a. a high pH (above 8).
    b. a pH between 4-6.
    c. phagocytic activity.
    d. a high loss rate.

24. A bacterium that is unique to the intestinal tract of breast-fed infants is
    a. *Bacteroides.*
    b. *Bifidobacterium bifidus.*
    c. *Escherichia coli.*
    d. Doderlein's bacillus.

25. In lichens, the fungi send projections of their hyphae across the algal cell wall in order to obtain nutrients from the photosynthetic partner. These projections are called
   a. mycelia.
   b. hyphal extensions.
   c. haustoria.
   d. lichenthropes.

26. The term ____________ is used to describe a mutually beneficial association in which the growth of an organism is dependent on one or more growth factors, nutrients, or substrates provided by another organism in the same vicinity.

27. Which of the following areas of the human body is (are) not normally free of microorganisms?
   a. respiratory tract
   b. kidneys
   c. eyes
   d. All of these are normally free of microorganisms.

28. The lungs are protected from microorganisms by
   a. the mucociliary blanket.
   b. lysozyme in mucus.
   c. phagocytic action of alveolar macrophages.
   d. All of these.

29. Establishing gnotobiotic colonies is much easier with mammals than with birds.

30.TF   The lower respiratory tract has a normal microbiota.

31. Microorganisms that are orally administered to promote health are called __________.

32.TF   The animals in gnotobiotic colonies are otherwise normal in all ways.

33.TF   The microbiota of the stomach is likely to reflect that of the oropharynx.

34.TF   Many anaerobic gram-negative bacteria are found in the duodenum.

35.TF   *Enterobacteriaceae* are found normally in the stomach.

36.TF   The colon has the largest microbial population in the body.

37.TF   *Escherichia coli* is the most prevalent bacterium found in the colon.

38.TF   The oral cavity becomes colonized with microorganisms within hours after birth.

39.TF   Yeasts are prevalent in the oral cavity.

40.TF   Many anaerobic organisms are present on the conjunctiva of the eye.

41.TF   The basic microbiota of the external ear resembles that of the skin.

42.TF   Some marine slugs engage in what can be called chloroplast symbiosis.

43.TF   The skin surface (epidermis) is a very favorable environment for colonization by microorganisms.

44.TF   The lower genitourinary tract is usually free of microorganisms.

45.TF   The main microorganisms found in the female reproductive tract are the acid-tolerant Lactobacillus species.

46.   Commensal microorganisms living on or in the skin are characterized as __________ if they multiply there.

47.   Commensal microorganisms living on or in the skin are characterized as __________ if they do not multiply there.

48.   Another term that has been used to describe syntropism is
a. cross-feeding.
b. the satellite phenomenon.
c. Both a and b are correct.
d. Neither a nor b is correct.

49.   The term that refers to those microbes found as symbionts in a short relationship of a nongrowing nature is __________.

50.   The term that refers to those microbes found as symbionts in a long relationship of a growing nature is __________.

51.   A close physical relationship between two different species that may or may not be beneficial is called __________.

52.   Bacteria that live within cells (the host) of another species are called __________.

53.   The chemolithotrophic bacteria that live in an endosymbiotic relationship with giant tube worms process __________ for energy.

54.TF   Male and female genitourinary tracts have nearly identical microbiota.

55.   __________ are capable of hydrolyzing the b'20 (1-4) linkages between successive D-glucose residues of cellulose.

56.   An animal in which all of the microorganisms present are known is called __________.

57.   In ruminants, the upper portion of the stomach expands to form a large pouch called the __________ and a smaller honey comb like __________.

58.TF Gnotobiotic animals can be but do not have to be, germ-free.

59.TF Although gnotobiotic animals are generally more susceptible to infection by microbial and other parasitic organisms, they are almost completely resistant to dysentery caused by *Entamoeba histolytica* because they lack the bacteria that this protozoan normally uses as a food source.

60.TF The ectosymbionts of ruminant animals are useful to their hosts in which of the following ways?
   a. They digest the cellulose of plant cell walls to provide access to the interior contents of the cells, which the ruminants can then metabolize.
   b. They produce most of the vitamins needed by the ruminant.
   c. They themselves are digested as nutrients when they enter the remaining stomachs of the ruminant.
   d. All of the above are correct.

Answers to Sample Test Questions from Video 10

1. Cellulase
2. Symbiosis
3. Mutualism
4. Fungi
5. Mycology
6. Methanotrophs
7. Fermentation
8. Methanogens
9. Mutualism
10. Biodegradation

Answers to Sample Test Questions from Text Chapter 25 (The Fungi [Eumycota], Slime Molds, and Water Molds) Web Site

1. False
2. Mycelium
3. True
4. Decomposer
5. Extracellular enzyme activity
6. Late blight of patato
7. Mycoses
8. True
9. Both plant roots and cyanobacteria
10. Terrestrial
11. False
12. True
13. True

Answers to Sample Test Questions from Text Chapter 28 (Symbiotic Associations: Commensalism, Mutualism, and the Normal Microbiota of the Human Body) Web Site

1. Symbiotic
2. Commensalism
3. True
4. Microflora on human skin
5. Bacteria
6. True
7. Brown algae
8. All of these

<table>
<tr><td>9. All of these</td><td>15. All of these</td></tr>
<tr><td>10. Any of these</td><td>16. False</td></tr>
<tr><td>11 In association with sebaceous glands</td><td>17. True</td></tr>
<tr><td>12. A pH between 4-6</td><td>18. False</td></tr>
<tr><td>13. Haustoria</td><td>19. True</td></tr>
<tr><td>14. Just a and b</td><td></td></tr>
</table>

Answers to Chapter 25 Sample Test Questions on The Fungi (Eumycota), Slime Molds, and Water Molds

The question in this test bank have been rated for difficulty. The ratings run from 1 (for the easiest) to 8 (for the most difficult).

#	Answer / Diff		#	Answer/Diff
1	D / 1		31	D / 1
2	D / 1		32	A / 5
3	B / 2		33	C / 5
4	A / 3		34	A / 8
5	C / 3		35	A / 7
6	D / 4		36	B / 3
7	D / 4		37	mycotoxin / 4
8	D / 5		38	C / 5
9	D / 5		39	C / 5
10	A / 6		40	B / 6
11	A / 5		41	F / 5
12	B / 3		42	T / 7
13	D / 2		43	T / 5
14	C / 4		44	T / 7
15	D / 2		45	F / 4
16	A / 7		46	T / 7
17	D / 5		47	*Ascomycota; Basidiomycota* / 7
18	C / 8		48	mycelium / 3
19	A / 3		49	chitin / 3
20	C / 5		50	yeast, fungal / 4
21	C / 4		51	decomposers / 6
22	A / 4		52	gametangium / 7
23	B / 5		53	mushroom / 6
24	C / 525 zoospores /5		54	myxomycetes/ 5
26	T / 6		55	1 A, 2 B, 3 C / 8
27	B / 728 B / 6		56	1 A, 2 B, 3 C / 8
29	B / 6		57	1 A, 2 B, 3 C / 8
30	B / 7		58	1 A, 2 B, 3 C / 8
			59	pseudoplasmodium / 7
			60	F / 5
			61	homothallic / 662
			62	heterothallic / 6

Answers to Chapter 28 Sample Test Questions on Symbiotic Associations: Commensalism, Mutualism, and the Normal Microbiota of the Human Body

The question in this test bank have been rated for difficulty. The ratings run from 1 (for the easiest) to 8 (for the most difficult).

#	Answer / Diff	#	Answer / Diff
1	D / 3	31	probiotics
2	B / 3	32	F / 4
3	A / 5	33	T / 4
4	D / 7	34	F / 4
5	ectosymbiosis / 6	35	F / 4
6	C / 6	36	T / 4
7	A / 5	37	F / 4
8	D / 8	38	T / 5
9	D / 3	39	T / 3
10	A / 6	40	F / 4
11	D / 7	41	T / 4
12	trophosome / 7	42	T / 2
13	D / 4	43	F / 5
14	D / 6	44	F / 3
15	D / 3	45	T / 2
16	C / 6	46	resident / 5
17	D / 4	47	transient / 5
18	D / 4	48	C / 5
19	D / 7	49	transient / 8
20	A / 7	50	resident / 8
21	B / 7	51	symbiosis / 8
22	C / 6	52	endosymbionts / 6
23	B / 4	53	hydrogen sulfide / 7
24	B / 6	54	F / 3
25	C / 8	55	Cellulases / 2
26	syntropism / 7	56	gnotobiotic / 4
27	C / 5	57	rumen; reticulum / 8
28	D / 4	58.	T / 5
29	F / 4	59.	T / 7
30	F / 5	60.	D / 6

Microorganisms and Humans
Program 11
Human Defenses

> **Pathogenicity** is not the rule. Indeed, it occurs so
> infrequently and involves such a relatively small number of
> species, considering the huge population of bacteria on Earth,
> that its has a freakish aspect. Disease usually results
> from inconclusive negotiations for symbiosis, an
> overstepping of the line by one side or the other, a biological
> misinterpretation of borders.
> – Lewis Thomas

One of microbiology's fifty most significant events in the past 125 years.

1882

In the above year, **Ilya Ilich (Elie) Metchnikoff** demonstrated that certain body cells move to damaged areas of the body, where they consume bacteria and other foreign particles. He called the process phagocytosis. He also proposed a theory of cellular immunity. With **Paul Ehrlich, Metchnikoff** was awarded the Nobel prize in Medicine or Physiology in 1908.

Student Assignments

- **Watch**   Video 11: *Human Defenses*
- **Read**   Chapter 29: Symbiotic Associations: Parasitism, Pathogenicity, and Resistance
               and
               Chapter 30: The Immune Response: Antigens and Antibodies in *Microbiology*, 4th ed., Prescott/Harley/Klein, McGraw-Hill
- **Lab Experiments** from *Laboratory Exercises in Microbiology*, 4th ed., Harley/Prescott, McGraw-Hill.
               Exercise 65: Isolation of Normal Microbiota from the Human Body
               Exercise 61: Agglutination II: Blood Groups
               Exercise 80: Demonstration of Koch's Postulates

Lesson Overview

Video 11: *Human Defenses*
Both nonspecific and specific defense strategies can defeat the invasion of microbial pathogens.
Here students learn about the coordinated defense system of humans through viral analogy,
animation, and examples of specific diseases.

Chapter 29: Symbiotic Associations: Parasitism, Pathogenicity, and Resistance
This chapter continues the discussion of symbiosis and focuses on parasitism. One of the possible
consequences of parasitism, pathogenicity, is also discussed. Most higher animals, including man,
possess various defense mechanisms that are employed against parasitic, pathogenic organisms.
Some of these are general, nonspecific mechanisms for resistance and are discussed in this chapter.
Other more specific mechanisms associated with the immune system are discussed in subsequent
chapters.

Chapter 30: The Immune Response: Antigens and Antibodies
This chapter introduces specific immunity, the defense mechanism that involves the recognition of
infectious agents, their products, some tumor cells, and certain macromolecules as foreign
materials. The body must produce specific defensive responses that will destroy or neutralize these
materials. This complex system involves several different types of responses and may lead to
several different problems of various system malfunctions. In this chapter, some aspects of the
immune system are introduced; other aspects are discussed in subsequent chapters.

Learning Objectives

After Successfully completing this unit, the student will be able to:

- discuss the general characteristics of parasitic symbiosis
- discuss the concepts of pathogens, disease, and infection
- describe the stages which pathogens must go through transmission, colonization,
  multiplication, and interference in order to cause disease
- discuss the general, physical, chemical, and biological barriers that organisms have for
  defending themselves against continual parasitic invasion
- explain the differences between specific and nonspecific immunity, artificial and natural
  immunity, and active and passive immunity
- describe the types of lymphocytes involved in immune responses and the different ways they
  respond to foreign substances
- describe the basic structure of antibody molecules and the different classes (isotypes) of these
  molecules
- describe the specificity and diversity of antibody molecules
- discuss the use of hybridomas to produce highly specific antibody molecules
- discuss the production and potential uses of catalytic antibodies

Course Integrator

Chapter 29
Host-Parasite Relationships (Parasitism) (p. 581)

Visual Resource Library	Symbiosis Figure 29.1
Microbes in Motion II	Microbial Pathogenesis/Principles of Pathogenicity
Web Link	*Microbiology* Online Learning Center
	http://www.mhhe.com/biosci/cellmicro/prescott/
Video *Unseen Life on Earth*	Unit #10 Host-Parasite Relations

Determinants of Infectious Disease (p. 582)

Visual Resource Library	Exotoxin Transport Mechanisms Figure 29.4
Transparency	Exotoxin Transport Mechanisms Figure 29.4
Visual Resource Library	Endotoxin Effects Figure 29.5
Microbes in Motion II	Microbial Pathogenesis/Adherence
	/Adherence Mechanisms, Receptors
Microbes in Motion II	Microbial Pathogenesis/Toxins/Exotoxins

General or Nonspecific Host Immune Defense Mechanisms (p. 591)

Transparency	Host Defenses Figure 29.6
Visual Resource Library	Effects of Fever in a Host Figure 29.7
Visual Resource Library	Antiviral Action of Interferon Figure 29.9
Transparency	Antiviral Action of Interferon Figure 29.9
Visual Resource Library	The RE System and Leukocytes Figure 29.10 & 11a
Transparency	The RE System and Leukocytes Figure 29.10 & 11a
Visual Resource Library	Effects of Acute Inflammation Figure 29.12
Visual Resource Library	Mast Cells and Inflammation Figure 29.13
Visual Resource Library	Phagocytosis Figure Number 29.14
Microbes in Motion II	Immunology/Innate (Non-specific) Immunity
	/Mechanical & Chemical Control
	/Cellular Defenses & Phagocytosis
Microbes in Motion II	Virology/Viral Pathogenesis/Host Cell Response
Microbes in Motion II	Immunology/Organs, Cells, and Cytokines
	/The Reticuloendothelial System

Chapter 30
Nonspecific Resistance (p. 607)

Visual Resource Library	Different Host Defenses Figure 30.1
	Microbes in Motion II Vaccines/Vaccine
	Development/Terminology, Composition
Microbes in Motion II	Vaccines/Immunization Schedules
Web Link	*Microbiology* Online Learning Center
	http://www.mhhe.com/biosci/cellmicro/prescott/
Video *Unseen Life on Earth*	Unit #11 Immune Response

Microbial Terms Used in Video 11 in Order of Appearance with Page Indexing to the Textbook (*Microbiology*, 4th ed., Prescott/Harley/Klein, McGraw-Hill)

skin 573
nonspecific resistance 591
immune response 606
sterile 136
toxins 284
chemotaxis 63
phagocytosis 601

macrophage 602
antigen 609
lymphocyte 607
plasma cell 620
memory cell 620
killer cell 608
cytotoxic T lymphocyte 608

T-killer cells 608
typhoid fever 803
acquired immunity 607
vacinations 607
vaccine 607
smallpox 744
polio 761
whooping cough 770

Web Links Related to Video 11 (Human Defenses) and Readings in Chapter 29 (Symbiotic Associations: Parasitism, Pathogenicity, and Resistance) and Chapter 30 (The Immune Response: Antigens and Antibodies) (Go to www.mhhe.com/prescott, Student on Line learning Center, Double Click on Chapter 29 [and then 30], Double Click on Web Sites

Chapter 29

Concepts of Fever. A review article by Philip Mackowiak, In *Archives of Internal Medicine*
http://www.ama-assn.org/sci-pubs/journals/archive/inte/vol_158/no_17/ira71067.htm

Microbial Virulence Factors
http://www.pedid.uthscsa.edu/001.htm

Understanding the Immune System. From the National Cancer Institute.
http://rex.nci.nih.gov/PATIENTS/INFO_TEACHER/immune_sys/Title.html

Chapter 30

Understanding the Immune System
http://rex.nci.nih.gov/PATIENTS/INFO_TEACHER/immune_sys/Title.html

Immune System Crossword Puzzle
http://fermat.stmarys-ca.edu/~jpolos/xwords/immune.html

Microbial Terms Used in Chapter 29 (Symbiotic Associations: Parasitism, Pathogenicity, and Resistance)

AM model 586
acute-phase reactant 592
adhesin 584
alpha hemolysis 591
antitoxin 586
bacterocin 597
beta hemolysis 591
beta-lysin 595
coagulation 590
colicin 597
colonization 584
cytotoxins 588
defensins 602
ectoparasite 581
endogenous pyrogen 591
endoparasite 581
endotoxin 589
endotoxin units 589
enterotoxins 588
exotoxin 586
fibronectin 595
final host 581
fomites 584

hemolysin 591
host 581
hyperferemia 592
hypoferemia 592
infection 582
infectious disease 582
infectious dose 582
infectivity 582
inflammation 597
integrins 598
interferon 595
intermediate host 581
intoxication 586
invasiveness 582
lethal dose 50 582
leukocidin 591
LPS-binding proteins 591
mucociliary blanket 594
neurotoxins 588
parasite 581
parasitic organism 581
parasitism 581

pathogen 582
pathogenicity 582
pathogenic potential 582
pathogenicity island 593
peristalsis 594
phagolysosome 601
phagosome 601
reservoir host 582
respiratory burst 602
selectins 597
septicemia 585
streptolysin-O 591
streptolysin-S 591
toxemia 586
toxigenicity 582
toxin 586
toxoid 586
transfer host 581
vectors 584
virulence 582
virulence factors 585

Microbial Terms Used in Chapter 30 (The Immune Response: Antigens and Antibodies)

acquired immunity 607
adjuvant 621
allotype 613
anamestic response 622
antibody 611
antibody-mediated immunity 609
antigen 610
antigen-binding fragment 611
antigenic determinant site 610
antiserum 621
artificially acquired active immunity 607
artificially acquired passive immunity 607
B cell 609
B lymphocyte 609
catalytic antibody 624
cell-mediate immunity 609
clonal selection theory 620
clone 620
colostrum 607
constant region 611
crystallizable fragment 611

cytokines 609
domain 611
epitope 610
GALT 616
hapten 610
humoral immunity 609
hybridoma 622
idiotype 613
IgA 615
IgD 615
IgE 615
IgG 615
IgM 615
immune response 606
immunity 606
immunogen 610
immunoglobulin 611
immunology 606
immunotoxin 623
isotype 613
J chain 614

lymphocytes 607
memory B cells 620
monoclonal antibody 622
myeloma cell 622
natural killer cell 609
naturally acquired active immunity 607
naturally acquired passive immunity 607
nonspecific resistance 607
opsonization 614
Peyer's patches 616
plasma cells 609
secretory IgA 615
serum 621
T cell 607
thymocyte 607
titer 621
T lymphocyte 607
vaccine 607
valence 610
variable region 611

Sample Test Questions from Video 11 (Human Defenses)

1. The human skin is home to millions of microorganisms that depend on our bodies for their own survival.
   a. True
   b. False

2. Harmful microbes can enter our bodies
    a. Though minor cuts.
    b. In food we eat.
    c. In the water we drink.
    d. All of these.

3. The skin is the __________ line of defense.
    a. first
    b. second
    c. third
    d. fourth

4. Where would you expect to find the most microbes on the human skin?
    a. On the back..
    b. On the scalp.
    c. On the sole of the foot.
    d. In an armpit.

5. The normal flora found on the skin can release chemicals that can destroy some potentially harmful bacteria.
    a. True
    b. False

6. The second line of defense is refereed to as __________.

7. Since phagocytes attack all invaders, this response is referred to as a specific response.
    a. True
    b. False

8. An example of a defensive cell that provides a link between the second and third lines of defense is the __________.

9. The third line of defense is a specific defense response.
    a. True
    b. False

10. The cells that produce antibodies are called __________.

11. Cells that are capable of living a long time and can produce antibodies are called __________.

12. T-killer cells can attach to cells infected with viruses and destroy them.
    a. True
    b. False

13. By introducing a vaccine, we are really just trying to trick our bodies into thinking it has seen a particular antigen before.
    a. True
    b. False

Sample Test Questions from Text Chapter 29 (Symbiotic Associations: Parasitism, Pathogenicity, and Resistance) Web Site

1. If a symbiont either harms or lives at the expense of another organism, the relationship is called
   - mutualism
   - symbiosis
   - parasitism
   - commensalism

3. An organism other than a human that is infected with a parasite that can also infect humans is called a
   - reservoir host
   - transfer host
   - intermediate host

4 . A short but severe infection is called
   - an acute infection
   - a chronic infection
   - a persistent infection

5. A condition of bacteria in the bloodstream that do not multiply is called bacteremia.
   - True
   - False

6. A latent infection
   - does not harm a healthy host
   - is one in which the organisms are present in tissue for a long time
   - develops during a hospital stay
   - is one in which more than one organism is present at the same time

7. A nosocomial infection
   - develops during a hospital stay
   - does not harm a healthy host
   - is one in which more than one organism is present at the same time

8. The final outcome of most host-parasite relationships is dependent on
   - the number of organisms present in or on the host
   - the virulence of the organism
   - the host's defenses
   - all of these

9. A lysozyme is an enzyme that lyses
   - O viruses
   - O fungi
   - O bacteria
   - O parasites

10. Fever augments the host's defenses by inhibiting the parasite's growth by raising the temperature above the optimum growth temperature.
    - O True
    - O False

11. Fever augments the host's defenses by decreasing the availability of iron to the organism.
    - O True
    - O False

12. Fever augments the host's defenses by stimulating leukocytes into action so that they can kill the organism.
    - O True
    - O False

13. A vector is an organism that transfers pathogens from one host to another.
    - O True
    - O False

Sample Test Questions from Text Chapter 30 (The Immune Response: Antigens and Antibodies) Web Site

1. The basic structure of the immunoglobulin molecule is
   - O four polypeptide chains: two heavy chains and two light chains
   - O four polypeptide chains: three heavy chains and one light chain
   - O two polypeptide chains: one heavy chain and one light chain

2. Two distinguishing features of immunoglobulins are their diversity and specificity.
   - O True
   - O False

3. Immunoglobulins specifically act with cells bearing antigens and mark them for subsequent removal.
   - O True
   - O False

4. Antibodies
   O   can destroy immunogens
   O   mark them for subsequent removal
   O   are part of the nonspecific immune response
   O   all of these
   O   can destroy the immunogen and mark them for subsequent removal

5. Acquired immunity
   O   develops after exposure to an antigen.
   O   is produced after antibodies are transferred from one individual to another
   O   can be obtained by natural or artificial means
   O   all of these

6. When an individual's immune system comes into contact with an antigenic stimulus during the
   O   course of daily activities, this is called
   O   naturally acquired passive immunity
   O   naturally acquired active immunity
   O   artificially acquired active immunity
   O   artificially acquired passive immunity

7. A vaccination is an example of
   O   naturally acquired passive immunity
   O   naturally acquired active immunity
   O   artificially acquired active immunity
   O   artificially acquired passive immunity

8. Which of the following cell types are produced in adult bone marrow?
   O   B cells
   O   T cells
   O   thymocytes
   O   all of these

9. Which of the following cell types constitute 20-30% of the circulating lymphocytes?
   O   B cells
   O   T cells
   O   thymocytes
   O   lymphocytes

10. T cells attack
   O   host cells that have been parasitized by microorganisms
   O   transplanted tissue cells from one host to another
   O   cancer cells
   O   all of these

11. The major immunoglobulin in human serum is
    - O IgG
    - O IgD
    - O IgM
    - O IgA

12. Log phase refers to the phase
    - O which directly follows a primary challenge with an antigen
    - O in which the antibody titer stabilizes
    - O in which the antibody titer rises quickly
    - O in which the antibodies are naturally metabolized

13. The immune system normally discriminates between ___________ antigens.
    - O self and nonself
    - O B and T cells
    - O humoral and cell-mediated
    - O primary and secondary response

14 Foreign substances to which lymphocytes respond are called antigens.
    - O True
    - O False

15. Each antigen has one antigenic determinant site or epitope.
    - O True
    - O False

16. Body temperature is a physiological barrier against certain microorganisms.
    - O True
    - O False

17. T cells are so named because they mature in the thyroid.
    - O True
    - O False

18. The type of immunoglobulin transferred from mother to baby during breast-feeding is
    - O IgA
    - O IgD
    - O IgG
    - O IgM

Chapter 29 Sample Test Questions on Symbiotic Associations: Parasitism, Pathogenicity, and Resistance

1. If a symbiont either harms or lives at the expense of another organism, it is called
   a. symbiosis.
   b. parasitism.
   c. commensalism.
   d. pathogenicity.

2. In parasitism, the host
   a. shelters the parasite.
   b. supports the growth of the parasite.
   c. is the smaller of the two organisms.
   d. shelters the parasite and supports the growth of the parasite.

3. Transfer of pathogens from host to environment and then to another host is called _________ contact.

4. Which of the following is an ectoparasite?
   a. horsehair worms
   b. ticks
   c. leeches
   d. All of these.

5. Molecules on the vascular endothelium to which neutrophils can attach during the inflammatory response are called
   a. adhesins.
   b. selectins.
   c. defensins.
   d. ligandins.

6. When the word parasite is used without qualification, it is referring specifically to
   a. nematodes.
   b. trematodes.
   c. cestodes.
   d. All of these.

7. The host on or in which the parasitic organisms either attain sexual maturity or reproduce is the
   a. intermediate host.
   b. transfer host.
   c. final host.
   d. reservoir host.

8. An organism other than a human that is infected with a parasitic organism that can also infect humans is called a(n)
   a. reservoir host.
   b. endoparasite.
   c. transfer host.
   d. intermediate host.

9. The abnormal condition in the host that results from invasion or establishment by pathogenic parasitic organisms in or on the host is called
   a. an infection.
   b. an infectious disease.
   c. pathogenesis.
   d. All of these.

10. Matching
    | 1. acute | a. Bacteria in the bloodstream that do not multiply |
    | 2. chronic | b. A short but severe course |
    | 3. bacteremia | c. Persisting over a long time |
    | 4. abscess | d. Small focus or area |

11. Matching
    | 1. latent | a. Caused during administration of health care |
    | 2. mass | b. Organisms present in tissue for long periods of time |
    | 3. iatrogenic | c. Not general |
    | 4. localized | d. A large number of pathogens in the circulation |

12. Matching
    | 1. mixed | a.. Does not harm a healthy host |
    | 2. overt | b. Develops during a hospital stay |
    | 3. opportunistic | c. More than one organism present at the same time |
    | 4. nosocomial | d. Symptomatic |

13. An infection caused by pus-producing organisms is called
    a. septic.
    b. pyogenic.
    c. systemic.
    d. toxemia.

14. An infection known as blood poisoning is called
    a. septicemia.
    b. systemic.
    c. toxemia.
    d. phytogenic.

15. The final outcome of most host-parasite relationships is dependent on
    a. the number of organisms present in or on the host.
    b. the virulence of the organism.
    c. the host's defenses.
    d. All of these.

16. Which of the following is not a requirement that a pathogen must possess in order to produce an infectious disease?
    a. initially be transported to the host
    b. be able to exit the host
    c. adhere to, colonize, or invade the host
    d possess the chemical ability to damage the host

17. An inanimate object that may be contaminated with a pathogen is called a
    a. vector.
    b. fomite.
    c. pathogen.
    d. All of these.

18. Which of the following is considered to be a biological defense mechanism?
    a. skin
    b. fever
    c. gastric acid
    d. indigenous microbiota

19. Which of the following is not used to help protect the lungs from infection?
    a. Turbulent airflow deposits airborne pathogens on sticky mucosal surfaces.
    b. The mucociliary blanket moves trapped organisms away from the lungs by ciliary action.
    c. Coughing and sneezing forcefully expel organisms away from the lungs.
    d. All of the above are used to protect the lungs from infection.

20. A subclass of fimbriae that bind procaryotes together for the transfer of genetic material are called
    a. fibrillae.
    b. capsules.
    c. pili.
    d. receptors.

21. Which of the following lyse erythrocytes causing anemia?
    a. hyaluronidase
    b. collagenase
    c. hemolysins
    d. coagulase

22. Which of the following clots fibrinogen in plasma?
    a. collagenase
    b. coagulase
    c. hemolysins
    d. hyaluronidase

23. Which of the following causes degranulation of lysosomes?
    a. hydrogen peroxide
    b. lecithinase
    c. leukocidins
    d. protein A

24. Which of the following is not a characteristic of an endotoxin?
    a. heat stable
    b. toxic at low doses
    c. weakly immunogenic
    d. generally the same regardless of source

25. Which of the following systems is activated by Hageman Factor?
    a. coagulation
    b. complement
    c. fibrinolytic
    d. All of these.

26. Streptolysin-S
    a. is produced by *Staphylococcus aureus*.
    b. causes alpha hemolysis.
    c. can act as a leukocidin.
    d. All of these.

27. The ability of some organisms to withhold iron from other organisms is called
    a. hypoferremia.
    b. hyperferremia.
    c. an acute-phase reaction.
    d. All of these.

28. Which of the following is not an exogenous pyrogen?
    a. gram-negative bacterial endotoxins
    b. N-acetylglucosamine-N-acetylmuramic acid
    c. soluble enterotoxin secreted by staphylococci
    d. iron

29. A fever augments the host's defenses by
    a. stimulating leukocytes into action.
    b. enhancing microbiostasis.
    c. enhancing the activity of the immune system.
    d. All of these.

30. Lysozyme is an enzyme that lyses
    a. viruses.
    b. fungi.
    c. bacteria.
    d. parasites.

31. Which of the following has no effect on the outcome of the host-parasite relationship?
    a. the number of parasites on/in the host
    b. the virulence of the parasite
    c. the defenses of the host
    d. All of these have an effect on the outcome of the host-parasite relationship.

32. Which of the following is not a way in which fever augments the host's defenses?
    a. inhibition of the parasite's growth by raising the temperature above
       the optimum growth temperature
    b. inhibition of growth by decreasing the availability of iron to the organism
    c. stimulation of leukocytes into action so that they can kill the organism
    d. All of these are ways that fever augments the host's defenses.

33. Which of the following is not a mode of action associated with endotoxin?
    a. shock
    b. paralysis
    c. diarrhea
    d. All of these are modes of action associated with endotoxin.

34. Phagocytosis leads to destruction of engulfed pathogens by which of the following mechanisms?
    a. lysosomal mediated hydrolysis
    b. phagosomal mediated respiratory burst
    c. both lysosomal mediated hydrolysis and phagosomal mediated respiratory burst
    d. neither lysosomal mediated hydrolysis nor phagosomal mediated respiratory burst

35. Under normal circumstances, which of the following is (are) normally sterile environments in mammals?
    a. kidneys
    b. ureters
    c. urinary bladder
    d. All of these.

36. Tears protect the eyes by
    a. flushing
    b. lysozyme
    c. flushing and lysozyme
    d. neither flushing nor lysozyme

37.TF Colonization specifically refers to the multiplication of a pathogen on or within a host.

38.TF   Fever response can be triggered either by an exogenous pyrogen released by the pathogen or by an endogenous pyrogen released by macrophages in response to the pathogen.

39.TF   Inflammation is one of the host's specific defense reactions to a tissue injury.

40.TF   Macrophages are phagocytic cells.

41.TF   The respiratory burst occurs in red blood cells.

42.TF   The respiratory burst increases ATP production.

43.TF   Fibronectin is a high molecular weight lipid that interacts with certain bacteria.

44.TF   Beta-lysin is a polypeptide released from blood platelets that can kill some gram-positive bacteria.

45.TF   The bacteriocins produced by *Escherichia coli* are called colicins.

46.TF   The respiratory system of mammals has formidable antiparasitic defense mechanisms.

47.TF   The mucociliary blanket is found in the digestive system of mammals.

48.TF   The short urethra in females is one reason why urinary tract infections are 14 times more common in females than in males.

49.TF   Cervical mucus has antibacterial activity.

50.   An __________ disease is a disease caused by the host's own microbiota because the host's resistance level is impaired.

51.   A(n) __________ is an organism that transfers pathogens from one host to another.

52.   The capacity of an organism to produce a toxin is called __________.

53.   __________ are soluble proteins produced by an organism as a result of its normal metabolism or its acquisition of a plasmid that redirects its metabolism.

54.   Exotoxins can be inactivated by iodine to form __________.

55.   A neutralizing antibody against a toxin is called a(n) __________.

56.   Because LPS is bound to the surface of bacteria, it is called a(n) __________.

57.   The term __________ refers to the degree or intensity of pathogenicity.

58.   Virulence is measured experimentally by either the __________ or __________.

59. Molecules known as __________ are broad spectrum antimicrobial peptides released from neutrophils.
    a. adhesins
    b. defensins
    c. integrins
    d. selectins

60.TF One of the usual symptoms of inflammation is a generalized fever.

61. Endotoxin is released from a pathogen when
    a. the microorganism whose cell wall contains the endotoxin lyses.
    b. the microorganism whose cell wall contains the toxin divides.
    c. Both a and b are correct.
    d. Neither a nor b is correct.

62TF Generally, exotoxins tend to be more heat stable than endotoxins.

Chapter 30 Sample Test Questions on The Immune Response: Antigens and Antibodies

1. Which of the following is not a phase in the immune response?
   a. recognition of the challenge
   b. amplification of the defense
   c. attack
   d. reattack

2. Which of the following refers to those general mechanisms that are inherited as part of the innate structure and function of each animal?
   a. nonspecific resistance
   b. nonspecific immunity
   c. specific immunity
   d. both nonspecific resistance and nonspecific immunity

3. Which of the following is a general barrier in the nonspecific defense of a mammalian host?
   a. inflammation
   b. phagocytosis
   c. fever
   d. mucous membranes

4. Which of the following is a biological barrier in the nonspecific defense of a mammalian host?
   a. mucous membranes
   b. inflammation
   c. interferons
   d. complement

5.   Antibodies
    a.   can destroy the immunogen.
    b.   can target the immunogen for destruction.
    c.   are part of the nonspecific immune response.
    d.   can destroy the immunogen and can target the immunogen for destruction.

6.   Acquired immunity refers to the type of specific immunity that
    a.   develops after exposure to suitable antigen.
    b.   is produced after antibodies are transferred from one individual to another.
    c.   can be obtained by natural or artificial means.
    d.   All of these.

7.   When an individual's immune system comes into contact with an appropriate antigenic stimulus during the course of daily activities, this is called
    a.   naturally acquired passive immunity.
    b.   naturally acquired active immunity.
    c.   artificially acquired active immunity.
    d.   artificially acquired passive immunity.

8.   The type of immunity that involves the transfer of antibodies from one host to another is called
    a.   naturally acquired passive immunity.
    b.   naturally acquired active immunity.
    c.   artificially acquired active immunity.
    d.   artificially acquired passive immunity.

9.   The transfer of antibodies in colostrum is a good example of
    a.   naturally acquired passive immunity.
    b.   naturally acquired active immunity.
    c.   artificially acquired active immunity.
    d.   artificially acquired passive immunity.

10.   A vaccination is a good example of
    a.   naturally acquired passive immunity.
    b.   naturally acquired active immunity.
    c.   artificially acquired active immunity.
    d.   artificially acquired passive immunity.

11.   Botulinum antitoxin produced in a horse and given to a human suffering from botulism food poisoning is an example of
    a.   naturally acquired passive immunity.
    b.   naturally acquired active immunity.
    c.   artificially acquired active immunity.
    d.   artificially acquired passive immunity.

12. Which of the following is not secondary lymphoid tissue?
    a. thymus
    b. bone marrow
    c. spleen
    d. lymph nodes

13. Which of the following cell types differentiate in the fetal liver and adult bone marrow?
    a. B cells
    b. T cells
    c. thymocytes
    d. All of these.

14. Which of the following cell types constitute 20-30% of the circulating lymphocytes?
    a. B cells
    b. T cells
    c. thymocytes
    d. lymphocytes

15. The general barriers that the body has for reducing infection are collectively referred to as
    a. nonspecific immunity.
    b. innate immunity.
    c. Both a and b are correct.
    d. Neither a nor b is correct.

16. Antibodies provide
    a. humoral immunity.
    b. cellular immunity.
    c. antibody-mediated immunity.
    d. both cellular immunity and antibody-mediated immunity.

17. The humoral immune response defends most against
    a. bacteria.
    b. bacterial toxins.
    c. viruses.
    d. All of these.

18. T cells attack
    a. host cells that have been parasitized by microorganisms.
    b. transplanted tissue cells from one host to another.
    c. cancer cells.
    d. All of these.

19. Matching
    1. IgG
    2. IgD
    3. IgM
    4. IgA

    a. Accounts for about 10% of the immunoglobin pool
    b. The major immunoglobin in human serum
    c. Has special features associated with secretory mucosal surfaces
    d. Found in trace amounts in the serum

20. Antibodies can be produced by
    a. natural means.
    b. immunizations.
    c. hybridomas.
    d. All of these.

21. Which of the following enhances the rate and quantity of antibody produced?
    a. adjuvant
    b. serum
    c. antiserum
    d. All of these.

22. Which of the following limits the use of antiserum?
    a. Antibodies obtained by this method are polyclonal.
    b. Hypersensitivity reactions can result from antiserum.
    c. Antiserum contains a mixture of antibodies.
    d. All of these limit the use of antiserum.

23. Matching
    1. log phase
    2. decline phase
    3. plateau phase
    4. lag phase

    a. This phase directly follows primary challenge with antiegn
    b. The antibody titer stabilizes.
    c. The antibody titer rises quickly.
    d. Antibodies are naturally metabolized.

24. Compared to the primary antibody response, the secondary antibody response
    a. has a shorter lag phase.
    b. has a more rapid log phase.
    c. persists for a longer plateau period.
    d. All of these.

25. Tumor cells of the immune system that produce large quantities of antibodies and can be readily cultivated are called
    a. myeloma cells.
    b. monoclonal antibodies.
    c. hybridomas.
    d. All of these.

26. An immunologic prevention of dental caries may soon be available if vaccination against caries-causing bacteria can build up sufficient specific __________ to be effective.
    a. IgG
    b. IgM
    c. IgA
    d. IgD

27. The secondary immune response is typically stronger than the primary response because __________ during the secondary response.
    a. all classes of immunoglobins are active
    b. both B and T cells are activated
    c. the antigen is weakened by the primary response
    d. a larger population of lymphocytes reactive to antigen has developed

28. The humoral immune response differs from the cell-mediated immune response in that only in the humoral immune response is there
    a. secretion of antibody.
    b. a precommitted lymphocyte.
    c. a clonal selection mechanism.
    d. the development of memory cells.

29. The immune system normally discriminates between __________ antigens.
    a. self and nonself
    b. B and T cells
    c. humoral and cell-mediated
    d. primary and secondary response

30. Which of the following diseases can be treated with passive immunization?
    a. infectious hepatitis
    b. cholera
    c. diphtheria
    d. typhoid

31. The constant region of an antibody molecule
    a. is responsible for its biological activity.
    b. enhances phagocytosis.
    c. activates complement.
    d. All of these.

32. All multicellular animals possess __________ serving as nonspecific immune mechanisms.
    a. anatomical barriers
    b. inflammatory responses
    c. physiological responses
    d. antibodies

33.   Natural killer cells specifically kill which of the following?
       a. tumor cells
       b. virus infected cells
       c. Both a and b are correct.
       d. Neither a nor b is correct.

34.   Which of the following types of immunoglobulins is most abundant in serum?
       a.   IgA
       b.   IgG
       c.   IgE
       d.   IgM

35.   Which of the following is not considered a benefit made possible by the use of monoclonal
       antibodies?
       a.   tissue typing for transplants
       b.   classification of T cell populations
       c.   targeted delivery of a chemotherapeutic agent
       d.   All of these are made possible by monoclonal antibodies.

36.   T cells produce and secrete proteins which do not directly interact with invading microorganisms
       but which augment the body's defense mechanisms. These molecules are called
       a.   antibodies
       b.   cytokines
       c.   immunogens
       d.   augmetins

37.   Which of the following is not a function of the Fc portion of the immunoglobulin molecule?
       a.   binding to host cells including some cells of the immune system and some phagocytic cells
       b.   binding to the first component of the complement system
       c.   binding to the antigen
       d.   All of these are functions of the Fc portion of the immunoglobulin molecule.

38.   The only immunoglobulin able to cross the placental barrier is
       a.   IgA
       b.   IgM
       c.   IgG
       d.   IgE

39.TF   Foreign substances to which lymphocytes respond are called antigens.

40.TF   Each antigen has one antigenic determinant site or epitope.

41.TF   Most antigens are monovalent.

42. Which of the following is a characteristic associated with secondary antibody responses?
a. shorter lag phase
b. higher antibody titer
c. higher antibody affinity
d. All of the above are correct.

43.TF Penicillin is a good example of a hapten.

44.TF Passive immunity cannot repeatedly be used successfully as a long-term treatment because injected antibodies will eventually trigger an immune response against the injected antibodies.

45.TF If the clonal selection theory is correct, there exists at least one lymphocyte capable of synthesizing an antibody specific to each antigen prior to exposure to the antigen.

46.TF Body temperature is a physiological barrier against certain microorganisms.

47.TF T cells are so named because they mature in the thyroid.

48.TF B cells are so named because, in birds, they mature in an appendage of the cloaca called the bursa of Fabricus.

49. In an antibody molecule, the stalk of the Y is called the __________.

50. In an antibody molecule, each loop along with approximately 25 amino acids on each side of the loop is called a __________.

51. In an antibody molecule, the constant domains of the heavy chain form the __________ region.

52. The change from low affinity antibody to high affinity antibody during a secondary immune response is referred to as
a. isotype switching
b. affinity maturation
c. anamnestic response
d. idiotype diversity

53. In an antibody molecule, the __________ region mediates binding to host tissue.

54. The ability of an antibody molecule to stimulate phagocytosis is called __________.

55. Combinatorial joining creates many combinations of __________ and __________ regions.

56. One unique property of antibodies is their remarkable __________.

57. A(n) __________ is an antibody coupled to a chemotherapeutic agent that can be used to target the delivery of the agent to a particular cell type.

58. A(n) ___________ is a small molecule that is not itself antigenic but is antigenic when bound to a larger molecule.

59. The type of immunoglobulin transferred from mother to infant during breast-feeding is
a. IgA
b. IgD
c. IgG
d. IgM

60. The specific regions within the variable region of an antibody that are responsible for antibody diversity and antigen specificity are called ___________ regions

Answers to Sample Test Questions from Video 11

1. A
2. D
3. A
4. D
5. A
6. Phagocytosis
7. B
8. Macrophage
9. A
10. B cells
11. Memory cells
12. A
13. A

Answers to Sample Test Questions from Text Chapter 29 (Symbiotic Associations: Parasitism, Pathogenicity, and Resistance) Web Site

1. Parasitism
2. Reservoir host
3. Reservoir host
4. An acute infection
5. True
6. Is one in which the organisms are present in tissue for a long time
7. Develops during a hospital stay
8. All of these
9. Bacteria
10. False
11. True
12. True
13. True

Answers to Sample Test Questions from Text Chapter 29 (The Immune Response: Antigens and Antibodies) Web Site

1. Four polypeptide chains
2. True
3. True
4. Can destroy immunogens and mark then for subsequent removal
5. Can be obtained by natural or artificial means
6. Naturally acquired active immunity

7. Artificially acquired active immunity
8. B cells
9. B cells

10. All of these
11. IgG
12. In which the antibody titer rises quickly
13. Self and nonself
14. True
15. False
16. True
17. False
18. IgA

Answers to Chapter 29 Sample Test Questions on Symbiotic Associations: Parasitism, Pathogenicity, and Resistance

The question in this test bank have been rated for difficulty. The ratings run from 1 (for the easiest) to 8 (for the most difficult).

#	Answer/Diff	#	Answer/Diff
1	B / 2	25	D / 6
2	D / 2	26	C / 8
3	indirect / 5	27	A / 7
4	D / 1	28	D / 7
5	B / 6	29	D / 6
6	D / 2	30	C / 5
7	C / 4	31	D / 5
8	A / 5	32	A / 8
9	A / 8	33	B / 6
10	1 C, 2 A, 3 B, 4 D / 5	34	C / 7
11	1 B, 2 D, 3 A, 4 C / 5	35	D / 4
12	1 C, 2 D, 3 B, 4 A / 5	36	C / 4
13	B / 7	37	T / 4
14	A / 6	38	T / 5
15	D / 6	39	F / 3
16	B / 7	40	T / 3
17	B / 4	41	F / 3
18	D / 5	42	T / 3
19	D / 5	43	F / 3
20	C / 7	44	T / 3
21	C / 6	45	T / 3
22	B / 6	46	T / 4
23	C / 6	47	F / 3
24	B / 7	48	T / 3

49	T / 3		56	endotoxin / 4
50	endogenous / 6		57	virulence / 6
51	vector / 4		58	lethal dose 50; infective dose 50 / 6
52	toxigenicity / 4		59	B / 7
53	Exotoxins / 5		60	F / 4
54	toxoids / 4		61	C / 6
55	antitoxin / 4		62	F / 5

Answers to Chapter 30 Sample Test Questions on The Immune Response: Antigens and Antibodies

The question in this test bank have been rated for difficulty. The ratings run from 1 (for the easiest) to 8 (for the most difficult).

#	Answer/Diff		#	Answer / Diff
1	D / 4		33	D / 6
2	D / 4		34	B / 4
3	C / 6		35	D / 5
4	B / 6		36	B / 6
5	D / 5		37	C / 5
6	D / 6		38	C / 7
7	B / 8		39	T / 2
8	A / 8		40	F / 5
9	A / 8		41	F / 6
10	C / 8		42	D / 4
11	D / 8		43	T / 4
12	B / 6		44	T / 8
13	A / 6		45	T / 8
14	A / 7		46	T / 3
15	D / 4		47	F / 5
16	D / 7		48	T / 5
17	D / 3		49	crystallizable fragment / 6
18	D / 4		50	domain / 7
19	1 B, 2 D, 3 A, 4 C / 19		51	constant / 7
20	D / 3		52	allotype / 7
21	A / 5		53	B / 6
22	D / 6		54	opsonization / 6
23	1 C, 2 D, 3 B, 4 A / 8		55	V; J / 5
24	D / 8		56	diversity / 5
25	A / 7		57	immunotoxin / 8
26	C / 8		58	hapten / 4
27	D / 8		59	A / 6
28	A / 8		60	hypervariable (complementarity determining) / 5
29	A / 8			
30	A / 8			
31	D / 7			
32	A / 6			

Microorganisms and Humans
Program 12
Microbes and Human Diseases

Infectious diseases constitute the major cause of death
worldwide and will not be conquered during our lifetimes.
–Emerging Infections

One of microbiology's fifty most significant events in the past 125 years.

1884

During the above year, **Robert Koch** published a *The Etiology of Tuberculosis*, in which he followed three steps: 1) the presence of the tubercle bacillus (as proved by staining) in tubercular lesions of various organs of humans and animals, 2) the cultivation of the organisms in pure culture in blood serum, and 3) the production of tuberculosis at will by its inoculation into guinea pigs. Koch was awarded the Nobel Prize in Physiology or Medicine in 1905.

Student Assignments

- **Watch**   Video 12: Microbes and Human Diseases
- **Read**   Chapter 35: The Epidemiology of Infectious Diseases in *Microbiology*, 4th ed., Prescott/Harley/Klein, McGraw-Hill.
- **Lab Experiments** from *Laboratory Exercises in Microbiology*, 4th ed., Harley/Prescott, McGraw-Hill.
  Exercise 80: Demonstration of Koch's Postulates

Lesson Overview

Video 12: *Microbes and Human Disease*
How microbes come into contact with humans, and the many factors leading to disease outbreaks around the globe, are examined here. Students learn about current efforts to track infectious diseases and the considerations necessary to control disease worldwide.

Chapter 35: The Epidemiology of Infectious Diseases
This chapter discusses the epidemiological parameters used to institute effective control, prevention, and eradication measures within an affected or potentially affected population. This chapter also discusses the epidemiology of hospital-acquired (nosocomial) infections, which have been of increasing concern in recent years.

Learning Objectives

After successfully completing this unit, the student will be able to:

- define *epidemiology* and explain how it relates to infectious diseases
- discuss the statistical parameters used to define and describe various infectious diseases
- discuss the need to identify the etiologic agent in order to trace the origin and manner of spread of an infectious disease outbreak
- describe the five epidemiological links (characteristics of the infectious organism, source and/or reservoir, mode of transmission, susceptibility of the host, and exit mechanisms) in the infectious disease cycle of any epidemic
- discuss the increase in nosocomial infections in recent years and the consequences of these infections

Course Integrator

Chapter 35

Microbes in Motion II   Epidemiology (entire book, especially the chapters on Terminology, Disease Acquisition, and Transmission)

Epidemiological Terminology (p. 722)
Web Link   *Microbiology* Online Learning Center
http://www.mhhe.com/biosci/cellmicro/prescott/

Measuring Frequency: The Tools of Epidemiologists (p. 722)

Infectious Disease Epidemiology (p. 723)
Video *Unseen Life on Earth*   Unit #12 Relationships: Microbes in Disease
Web Link   Weekly Epidemiological Record:
http://www.who.int/wer/
Web Link   Centers for Disease Control Homepage
(http://www.cdc.gov/)

Recognition of an Infectious Disease in a Population (p. 724)

Recognition of an Epidemic (p. 724)

Visual Resource Library	Epidemic Curves Figure 35.3
Transparency	Epidemic Curves Figure 35.3
Visual Resource Library	Herd Immunity Figure 35.5
Transparency	Herd Immunity Figure 35.5

The Infectious Disease Cycle (p. 727)

Virulence and the Mode of Transmission (p. 731)

The Emergence of New Diseases (p. 732)

Control of Epidemics (p. 732)

Nosocomial Infections (p. 734)

Microbial Terms Used in Video 12 in Order of Appearance with Page Indexing to the Textbook (*Microbiology* 4th ed., Prescott/Harley/Klein, McGraw-Hill)

Giardia lamblia 832	Ebola 745	cytokines 627
infectious disease 582	1918 influenza 740	plague 777
pathogens 727	HIV 749	anthrax 777
Neisseria 780	Legionaires disease	four corner's disease
Candidia 819	768	732
Salmonella 724	Hanta virus 732	

Web Links Related to Video 12 (Microbes and Human Diseases) and Readings in Chapter 35 (The Epidemiology of Infectious Disease) (Go to www.mhhe.com/prescott, Student On Line Learning Center, Double Click on Chapter 35, Double Click on Web Sites)

Centers for Disease Control Home Page
http://www.cdc.gov/

Morbidity and Mortality Weekly Report. The data in the weekly MMWR are provisional, based on weekly reports to the Centers for Disease Control and Prevention by state health departments. The reporting concludes at close of business on Friday; compiled data on a national basis are officially released to the public on the succeeding Friday.
http://www2.cdc.gov/mmwr/

Emerging Infectious Diseases. Tracking trends and analyzing new and emerging infectious diseases around the world from CDC.
http://www.cdc.gov/ncidod/EID/index.htm

World Health Organization Home Page
http://www.who.int/

*Weekly Epidemiological Record (*from the WHO). The Weekly Epidemiological Record (WER) serves as an essential instrument for the rapid and accurate dissemination of epidemiological information on cases and outbreaks of diseases under the International health Regulations and on other communicable diseases of public health importance, including the newly emerging or re-emerging infections.
http://www.who.int/wer/

Microbial Terms Used in Chapter 35 (The Epidemiology of Infectious Disease)

active carrier 727
acute carrier 727
airborne transmission 728
antigenic drift 725
antigenic shift 725
autogenous infection 735
biologic transmission 731
carrier 727
chronic carrier 727
common-source epidemic 724
common vehicle transmission 731
communicable disease 727
contact transmission 729
convalescent carrier 727
disease 721
disease syndrome 724
droplet nuclei 728
endemic disease 722

enzootic 722
epidemic 722
epidemiologist 721
epidemiology 721
epizootic 722
epizootiology 722
fomite 731
harborage transmission 731
health 721
healthy carrier 727
herd immunity 724
hyperendemic disease 722
incubation period 724
incubatory carrier 727
index case 722
infectious disease cycle 727
morbidity rate 722
mortality rate 723
nosocomial infection 734

outbreak 722
pandemic 722
panzootic 722
period of infectivity 727
prevalene rate 723
prodromal stage 724
propagated epidemic 724
puerperal fever 730
reservoir 727
signs 723
source 727
sporadic disease 722
statistics 722
symptoms 724
transient carrier 727
vector 727
vector-borne transmission 731
vehicle 731
zoonoses 722

Sample Test Questions from Video 12 (Microbes and Human Disease)

1. Infectious diseases are as wide ranging as the microbes that cause them.
   a. True
   b. False

2. In 1918, 20 million people died in two years as a deadly strain of __________ swept the world.
   a. Measles
   b. Smallpox
   c. Influenza
   d. Cholera
   e. Plague

3. The discovery of __________ heralded the age of antibiotics.

4. The natural host for the Hanta fever virus is a __________.

5. Initially, the hanta virus penetrates the endothelial cells of the capillaries of the __________ causing pulmonary edema.

6. The English Sweating Sickness is caused by a virus similar to the Hanta virus in New Mexico.
   a. True
   b. False

7. The Hanta virus is an
   a. emerging virus
   b. re-emerging virus

8. Recently researchers have discovered a strain of hanta virus that seems to move from human to human.
   a. True
   b. False

9. Hanta virus has been kept in check because contact with a particular rodent's feces was required in order to become infected.
   a. True
   b. False

10. Currently, the hanta virus is among the least deadly of all known viruses.
    a. True
    b. False

Sample Test Questions from Text Chapter 35 (The Epidemiology of Infectious Disease) Web Site

1. The science of epidemiology deals with the occurrence and distribution of disease in a population.
   - O   True
   - O   False

2. When a disease maintains a relatively steady, low-level it is called
   - O   an epidemic.
   - O   hyperendemic
   - O   endemic
   - O   sporadic

3. A sudden increase in the occurrence of a disease above the expected level is called
   - O   sporadic
   - O   hyperendemic
   - O   an epidemic

4. A pandemic is a long-term increase in disease occurrence within a large population.
   - O   True
   - O   False

5. Moderate prevalence of a disease in animals is called
   - O   enzootic
   - O   epizootic
   - O   zoonotic

6. Diseases of animals that can be transmitted to humans are called
   - O   animal and human diseases
   - O   zoonoses
   - O   communicable diseases

7. A(n) __________ rate measures the number of individuals who become ill as a result of a specific disease within a susceptible population during a specific time period.
   - O   epidemic
   - O   morbidity
   - O   mortality
   - O   index

8. The prevalence rate refers to the total number of individuals infected at any one time regardless of when the disease began.
   - O   True
   - O   False

9. The _____________ rate is the relationship of the number of deaths from a disease to the total number of cases.
   O   prevalence
   O   morbidity
   O   mortality

10. Which type of immunity explains the resistance of a population to infection?
    O   herd
    O   innate
    O   natural

11. Why do epidemics of infectious organisms continue to occur?
    O   new strains of organisms continue to evolve
    O   new organisms continue to evolve
    O   new organisms and new strains continue to evolve

12. The first link in the infectious disease cycle is the
    O   source
    O   reservoir
    O   pathogen
    O   transmission

13. An organism that is able to be transmitted from one host to another is described as infective.
    O   virulent
    O   pathogenic
    O   communicable

14. The natural location in which the pathogen is normally found is called the
    O   reservoir
    O   source
    O   carrier

Chapter 35 Sample Test Questions on the Epidemiology of Infectious Disease

1.   The science of epidemiology
     a.   evaluates the control of disease in a defined population.
     b.   evolved in relation to the great epidemic diseases.
     c.   embraces all diseases.
     d.   All of these.

2. When a disease maintains a relatively steady, low-level frequency at a moderately regular interval, it is called
   a. sporadic.
   b. hyperendemic.
   c. endemic.
   d. an epidemic.

3. The sudden increase in the occurrence of a disease above the expected level is called
   a. sporadic.
   b. hyperendemic.
   c. endemic.
   d. an epidemic.

4. The first case in an epidemic is called the
   a. break case.
   b. index case.
   c. first case.
   d. All of these.

5. A(n) __________ is a long-term increase in disease occurrence within a large population.
   a. endemic
   b. pandemic
   c. epidemic
   d. sporadic disease

6. Moderate prevalence of a disease in animals is called
   a. enzootic.
   b. epizootic.
   c. panzootic.
   d. zoonotic.

7. Diseases of animals that can be transmitted to humans are called
   a. animal and human diseases.
   b. zoonoses.
   c. communicable diseases.
   d. contagious diseases.

8. A __________ rate measures the number of individuals who become ill as a result of a specific disease within a susceptible population during a specific time period.
   a. prevalence
   b. morbidity
   c. mortality
   d. index

9. The __________ rate refers to the total number of individuals infected at any one time regardless of when the disease began.
   a. prevalence
   b. morbidity
   c. mortality
   d. index

10. The __________ rate is the relationship of the number of deaths from a given disease to the total number of cases of the disease.
   a. prevalence
   b. morbidity
   c. mortality
   d. index

11. An infectious disease
   a. may be mild.
   b. may be severe.
   c. may cause death of the host.
   d. All of the above are correct.

12. A __________ epidemic is characterized by a sharp rise to a peak then a rapid, but not as pronounced, decline in the number of individuals infected.
   a. propagated
   b. common-source
   c. herd
   d. All of these.

13. A __________ epidemic is characterized by a relatively slow and prolonged rise then a gradual decline in the number of individuals infected.
   a. propagated
   b. common-source
   c. herd
   d. All of these.

14. Which type of epidemic usually results from a single contaminated source such as food?
   a. propagated
   b. common-source
   c. herd
   d. All of these.

15. Which type of epidemic usually results from the introduction of a single infected individual into a susceptible population?
   a. propagated
   b. common-source
   c. herd
   d. All of these.

16. Which type of immunity is an epidemiological concept that explains the resistance of a population to infection and the spread of an infectious organism due to the immunity of a high percentage of the population?
    a. natural
    b. experimental
    c. herd
    d. innate

17. Which of the following is (are) reasons why epidemics of infectious organisms continue to occur?
    a. continuous evolution of new strains of an organism
    b. continuous evolution of new organisms
    c. Both a and b are correct.
    d. Neither a nor b is correct.

18. The first link in the infectious disease cycle is the
    a. source.
    b. reservoir.
    c. organism.
    d. transmission.

19. An organism that is able to be transmitted from one host to another is said to be
    a. infective.
    b. virulent.
    c. pathogenic.
    d. communicable.

20. The site or natural environmental location in which the causative organism is normally ound is called the
    a. reservoir.
    b. source.
    c. carrier.
    d. animate phase.

21. An infected individual who is a potential source of infection for others and plays an important role in the epidemiology of the disease is a
    a. carrier.
    b. reservoir.
    c. host.
    d. All of these.

22. A(n) __________ carrier is an individual who is incubating the organism in large numbers but is not yet ill.
    a. active
    b. convalescent
    c. healthy
    d. incubatory

23. A(n) __________ carrier is an individual who has an overt clinical case of a disease.
    a. active
    b. convalescent
    c. healthy
    d. incubatory

24. A(n) __________ carrier is an individual who harbors the infectious organism but is not ill.
    a. active
    b. convalescent
    c. healthy
    d. incubatory

25. Individuals who harbor an organism for long periods are called
    a. chronic carriers.
    b. causal carriers.
    c. acute carriers.
    d. transient carriers.

26. In humans, the most common source of zoonoses are
    a. wild animals.
    b. domestic animals.
    c. insects.
    d. ticks.

27. In which type of transmission is the organism truly suspended and travels over a meter or more from the source to the host?
    a. contact
    b. airborne
    c. vehicle
    d. vector-borne

28. The small particles that represent what is left from the evaporation of larger particles are called
    a. droplets.
    b. droplet nuclei.
    c. infectious dust.
    d. All of these.

29. The material that settles out of the air then becomes resuspended as a result of physical forces is called
    a. droplets.
    b. droplet nuclei.
    c. infectious dust.
    d. an animate vector.

30.	Contact transmission can be
	a.	direct.
	b.	indirect.
	c.	droplet.
	d.	All of these.

31.	___________ contact refers to the transmission of an infectious organism from the source to the host via an intermediary.
	a.	Direct
	b.	Person-to-person
	c.	Indirect
	d.	Droplet

32.	___________ contact implies a coming together or touching between the source or reservoir of the infectious disease organism and the host.
	a.	Direct
	b.	Person-to-person
	c.	Indirect
	d.	Droplet

33.	Which of the following is (are) the major concern(s) of epidemiologists?
	a.	the discovery of factors essential to disease occurrence
	b.	the development of methods for disease prevention
	c.	Both of these answers are correct.
	d.	Neither of these answers is correct.

34.	Which of the following is not a mechanism by which new susceptible individuals enter a population?
	a.	birth of new individuals
	b.	migration of individuals into the population
	c.	evolution of disease causing organisms such that they are no longer recognized by host immune mechanisms
	d.	All of these are mechanisms by which new susceptible individuals enter a population.

35.	Which of the following is not normally a mechanism of introduction of airborne organisms into the air from humans or animals?
	a.	sneezing
	b.	evaporation of sweat
	c.	coughing
	d.	vocalization (talking)

36.TF	An infection that is incubating in a patient at the time of admission and which develops clinical symptoms during the patient's stay in a health care facility is not considered nosocomial.

37. Chlorination of water supplies represents an epidemiological technique directed at
    a. eliminating susceptible individuals.
    b. eliminating the source of infection.
    c. eliminating both susceptible individuals and the source of infection.
    d. eliminating neither susceptible individuals nor the source of infection.

38.TF Epidemiologists can recognize an infectious disease in a population by utilizing various surveillance methods.

39.TF After an infectious disease has been recognized in a population, the next step is to correlate the disease outbreak with a specific organism.

40.TF The clinical microbiology laboratory has as its main purpose conducting population surveys to determine who has a certain disease.

41.TF The clinical microbiology laboratory may assist in the eradication of a disease.

42.TF Quarantine of carriers represents an epidemiological technique directed at reducing the source of infection.

43.TF Active immunization of a population will raise the general level of herd immunity.

44.TF Nosocomial infections are ones that develop within any health care facility.

45.TF Most organisms that cause nosocomial infections come from endogenous sources.

46.TF All personnel involved with the care of patients should be familiar with basic infection control measures.

47. Inanimate objects involved in the transmission of an infectious organism are called __________.

48. In epidemiology, common vehicles involved in the transmission of an infectious organism are called __________.

49. Living transmitters of an infectious organism are called __________.

50. __________ transmission implies that the organism goes through some morphological or physiological change within the vector.

51. The fourth link in the infectious disease cycle is the __________.

52. The last link in the infectious disease cycle is the __________ of the infectious organism from the host.

53. __________ escape is best exemplified by movement of the organism from the host via feces, urine, or saliva.

54. The cause of a disease is referred to as the disease __________.

55. Which of the following is not used to reduce or eliminate the source of an infection?
a. treatment of sewage to reduce water contamination
b. destruction of vectors by spraying insecticides
c. destruction of an animal reservoir of the infectious agent
d. All of the above are used to reduce or eliminate the source of an infection.

56. The immediate location from which a disease-causing organism is transmitted to a host either directly or through an intermediate is called the __________.

57. The location where a disease-causing organism is normally found is called the __________.

58.TF Pathogens that are spread by direct contact tend to be more virulent than those that are vector-borne.

59. Which of the following contributes to the emergence of new diseases?
a. Rapid transportation systems and the mobility of the population
b. Ecological disruption such as the loss of predators and the destruction of rain forests
c. Increased drug usage and sexual promiscuity
d. All of the above.

60. A set of signs and symptoms that are characteristic of a particular disease is called a __________.

61. Objective changes in the body that can be observed or measured are called __________, while subjective changes experienced by the patient are called __________.

62. The period, during the cause of a disease, in which the signs and symptoms begin but are not yet clear enough for a diagnosis is called the __________ period.

63. Which of the following contributes to the emergence of new diseases?
a. rapid transportation systems and the mobility of the population
b. ecological disruption such as loss of predators and/or loss of habitat
c. increased drug usage and sexual promiscuity
d. All of the above can contribute to the emergence of new diseases.

64. Which of the following is not considered a newly emergent disease?
a. *Escherichia coli* O157:H7
b. cryptosporidiosis
c. hantavirus pulmonary syndrome
d. All of the above are considered newly emergent diseases.

Answers to Sample Test Questions from Video 12

1. A	6. True
2. C	7. B
3. Penicillin	8. B
4. Deer mouse	9. A
5. Lungs	10. B

Answers to Sample Test Questions from Text Chapter 35 (The Epidemiology of Infectious Disease) Web Site

1. True	8. True
2. Endemic	9. Mortality
3. An epidemic	10. Herd
4. True	11. New organisms and new strains
5. Enzootic	12. Pathogen
6. Zoonoses	13. Communicable
7. Morbidity	14. Reservoir

Answers to Chapter 35 Sample test Questions on the Epidemiology of Infectious Disease

The question in this test bank have been rated for difficulty. The ratings run from 1 (for the easiest) to 8 (for the most difficult).

#	Answer / Diff	#	Answer / Diff
1	D / 2	18	C / 5
2	C / 5	19	D / 4
3	D / 5	20	A / 5
4	B / 4	21	A / 6
5	B / 4	22	D / 6
6	A / 4	23	A / 6
7	B / 4	24	C / 6
8	B / 5	25	A / 6
9	A / 5	26	B / 3
10	C / 5	27	B / 5
11	D / 3	28	B / 5
12	B4	29	C / 6
13	A / 4	30	D / 6
14	B / 4	31	C / 6
15	A / 4	32	A / 4
16	C / 4		
17	C / 7		

33	C / 2		50	Biologic / 5
34	D / 4		51	host / 5
35	B / 5		52	exit / 5
36	T / 6		53	Passive / 5
37	B / 4		54	etiology / 5
38	T / 2		55	B / 8
39	T / 2		56	source / 5
40	F / 2		57	reservoir / 5
41	T / 2		58	F / 8
42	T / 3		59	D / 8
43	T / 3		60	syndrome / 6
44	T / 3		61	signs; symptoms / 6
45	F / 3		62	prodromal / 6
46	T / 3		63	D / 6
47	vehicles / 5		64	D / 4
48	fomites / 5			
49	vectors / 5			